Darwinian Deceptions

Darwinian Deceptions

Defending Truth for Today's Latter-Day Saints

Dave S. Collingridge, PhD (Editor)
David M. Barker, CPA
Ryan Nilsson, CPA
Eric N. Skousen, PhD

Published by Digital Legend Press,
Salt Lake City, UT

(Author royalties from the sale of this book are donated to the Church General Missionary Fund)

Darwinian Deceptions compiled and authored by Dave S. Collingridge, PhD (Editor). Contributors: David M Barker, CPA; Ryan Nilsson, CPA; Eric N. Skousen, PhD.

Published by Digital Legend Press, a division of Legends Library Publishing, Inc. 1994 Forest Bend Drive, Salt Lake City UT 84121 (www.digitalegend.com).

For information about special discounts available for bulk purchases, sales promotions, contact Legends Library Publishing at 877-222-1960 or info@digitalegend.com.

ISBN: 978-1-944200-48-0

Cover design by Boyd J Tuttle - Interior layout by Dave S. Collingridge

First Printing March 2018. Printed in USA.

Contents

Chapter 1

Preface

Darwin Week at BYU

The year 2009 was a special time for evolutionists. It marked Charles Darwin's 200th birthday. I was teaching part-time on the BYU campus in Provo, Utah. I remember seeing books, articles, and speeches honoring his achievements.

The BYU biology department celebrated all week with free public lectures on Darwin. Curious about what sort of things were being presented by the biology department, I obtained a copy of the lecture schedule.

My attention was drawn to a lecture on the similarities between Joseph Smith, Abraham Lincoln, and Charles Darwin. What similarities could be made about the prophet of the Restoration, a great American president, and the man who popularized a controversial theory on the origins of life? I wanted to find out, so I decided to attend.

Shortly after I arrived, the seats in the large auditorium were full. Students who could not find a seat were scattered along the outside edges of the auditorium. Some were standing. Some were sitting on the floor. By the time the lecturer got up to speak, the auditorium was filled beyond capacity.

In a small way, I anticipated a loudmouth defender of

Darwinism to stand up and, with evangelical fervor, tell us all the reasons why we should accept Darwinism. This did not happen. The lecturer was the opposite of a firebrand evolutionist – modest and pleasant. His demeanor assured me that I was not about to receive a tongue lashing about how I am scientifically ignorant if I don't accept all of evolution, something I had been accused of before.

During the lecture I reflected on some of the comparisons being made between Joseph Smith, Abraham Lincoln, and Charles Darwin. Other than the fact that they were born within four years of each other, the comparisons seemed strange. I thought it unusual to consider Charles Darwin's contributions alongside those of Abraham Lincoln and Joseph Smith. Comparing Darwin's contributions to those of Joseph Smith and Abraham Lincoln is a bit like comparing a peddle bike to a Harley Davidson motorcycle; both have a seat, two wheels, and two handle bars. Other than that, not much else is the same.

As the lecture came to a close, I had no idea the most educational part of the evening was yet to come. Immediately after the speaker finished, the crowd erupted into an enthusiastic applause. Some were standing. Some were cheering and hooting. I looked around in disbelief. The energetic applause caught me off guard. For a second I wondered if we had heard the same lecture. I thought the lecture was fine. They thought it was great! Not wanting to look out of place, I added a few extra claps to my subdued applause.

Walking back to my car, I wondered at what I had just witnessed. I saw a large crowd of BYU students energetically approve of a lecture on the similarities between Darwin and two eminent men in American history. I also reflected on all the posters and articles circulating around campus announcing Darwin Week. Judging by the praise being heaped upon Darwin all that week, it was clear to me that the man is held in high regard. Unbeknownst to me, Darwin had become a pop star.

Praising Darwin

I don't have a problem with honoring Darwin's positive achievements. His detailed descriptions of creatures certainly expanded our appreciation for the variety of life. There is also the positive impact his discovery of natural selection has had on our understanding of how different characteristics arise within species. He discovered that organisms possessing characteristics which increase the chances of survival have an increased likelihood of surviving and passing on those same characteristics to their offspring. Survival characteristics that are passed on may eventually become dominant within a species.

However, I do have a problem with Darwin becoming a pop star. When scientists are granted hero status, people are less likely to challenge their ideas. Their ideas are considered unassailable. (The history of science contains several examples of this phenomenon.)

Unassailable is an accurate description of evolution's standing today. We are bombarded with messages from schools, books, television, and the internet suggesting that Darwin was a great scientist and that his ideas are beyond reproach. This messaging is apparent whenever we see a news report about some newly discovered evidence related to our supposed primate heritage. According to these types of reports, it is reasonable to question what the evidence says about our assumed primate heritage, but it is not reasonable to question whether our ancestors were primates in the first place. Supposedly the idea that we evolved from lower primates has already been proven.

Spiritual Harm

Turning Darwin into a pop star would not be so bad if some of his ideas were not so damaging, in a spiritual sense. Unfortunately, spiritually damaging ideas were evident during the Darwin celebrations at BYU.

While walking across campus during Darwin Week, I picked up a copy of the university newspaper, the Daily Universe. The newspaper contained several articles and announcements related to Darwin Week festivities. As usual, I looked for the letters to the editor first—a favorite of mine for morally opinionated content—and then moved onto the articles. I didn't have to read long before I found what I was watching for—a statement from a faculty member advocating a dangerous evolutionary view. The Daily Universe quoted a BYU biology faculty member as saying, "We [the biology department] spend time dispelling the myth that evolution and religion are incompatible" (available at http://newsnet.byu.edu/pdf/du20090209.pdf).

How can saying "We spend time dispelling the myth that evolution and religion are incompatible" be spiritually harmful? To accept that evolution and the gospel are compatible would suggest that their accounts of the creation of humanity are compatible. Yet they are not. According to evolution, humans evolved from lower life forms through blind, undirected natural processes. However, according to the scriptures, our first parents, Adam and Eve, were created by God. The scriptures also teach that the creation of Adam and Eve was directed and purposeful, not random and haphazard. Evolutionary ideas that contradict these revealed truths have the potential to create confusion and cause people to question their religious beliefs.

Something else in that statement caught my attention. By advocating full compatibility between the gospel and evolution, I suspected the biology department was encouraging students to accept evolutionary views that contradict their gospel beliefs. My suspicion was confirmed the following day.

A Gospel-Friendly Perspective

The next day while teaching about evolutionary psychology in my History of Psychology class, I decided to give students a brief critique of the theory of evolution. Amid all the celebrations and praise being heaped on Darwin that week, I wanted students to hear an alternative viewpoint. I wanted them to hear theological concerns with reconciling the gospel and evolution. I also wanted them to know it is scientifically reasonable to reject portions of evolutionary theory. From what I could tell, these sorts of ideas were not being expressed during Darwin Week.

I started my critique by sharing a few of evolution's strengths. These strengths included ideas which have been proven true, such as that a certain amount of evolution occurs within species. Evolution within species, sometimes called microevolution, adds variety to life forms of the same kind. We also know that by studying evolution among microbes such as bacteria and viruses, we can enhance our understanding of antimicrobial resistance and possibly develop better antibiotics and antiviral medications.

Next, I talked about the weaknesses of evolution. I told the class that the idea that life can evolve from one form into a completely different form has not been demonstrated. Yet what do we repeatedly hear from evolutionists? We hear that evolution across life forms has been proven beyond a reasonable doubt.

Because this was a BYU class, I also talked about evolutionary beliefs that contradict gospel teachings. I talked about some of the problems that arise when people attempt to reconcile the gospel and evolution. I told the students that in order to reconcile the two, beliefs in either one or the other must be altered to bring them into harmony. Far too often spiritual beliefs are altered to make religion fit with evolution.

Following my critique, I encouraged students to discuss the issues. As is often the case in a large university class, many of the students remained silent and kept their thoughts to themselves. Some of the students expressed agreement with the ideas I presented. A few others openly disagreed. Overall it was a thought-provoking discussion.

When class was finished, I picked up my notes and headed for the door. As I was leaving, a young woman approached and asked if we could talk. Thinking she might want to say something in private, I invited her to walk outside with me.

As we exited the Spencer W. Kimball tower, she expressed gratitude for the things I said about evolution. She was particularly thankful I had shared a perspective on evolution that stayed true to the gospel. I asked her why she felt that way. She said she was feeling pressured by her biology instructor to accept evolutionary ideas that contradict gospel teachings.

I assured her it is completely reasonable to reject evolutionary ideas that contradict gospel truths. True science will always agree with true religion, but when they do not agree, science should be questioned, not revealed truth. Gospel truths are eternal. Scientific beliefs, on the other hand, are continually changing.

My discussion with this faithful Latter-day Saint confirmed my expectation: Some members of the Church are feeling pressured to accept evolutionary ideas that oppose their religious beliefs. I also learned something else. Latter-day Saints are looking for perspectives on evolution that stay true to gospel teachings. Unfortunately, there are few resources offering a balanced, gospel-friendly perspective on evolution.

What this Book Offers

This book presents a well-reasoned perspective on evolution that stays true to the plain and simple truths found in the gospel of Jesus Christ. Readers will discover it is completely reasonable to reject evolutionary concepts that contradict gospel truths.

This book was written for Latter-day Saints who are curious about what to believe when it comes to evolution. They may be searching for answers to questions such as: What parts of evolution are compatible with the gospel? How does evolution differ from gospel teachings on the creation? Has evolution really been proven true? And what have the prophets and apostles said about evolution? Answers to these and many more questions are found within these pages.

This book was also written for members of the Church who are feeling pressured to accept evolution. Tired of seeing evolutionary messages that contradict their religious beliefs, many members are looking for gospel-friendly perspectives on the subject. They are looking for material that challenges evolutionary dogma while staying true to their religious beliefs. This book fulfills this purpose. It presents scientific discoveries and secular insights that oppose Darwinian thinking, and it does so without compromising the plain and simple truths of the gospel.

Unless otherwise stated, whenever we use the words "evolution" and "Darwinism," we are referring to the idea that life can change into a completely different life form, and that animals and humans evolved from a common ancestor. As we critique these beliefs from a rational-empirical and gospel-themed perspective, readers will discover that evolution is not sacrosanct. Like every other scientific theory, it is subject to criticism and refutation. These criticisms and refutations are especially evident within the light of the restored gospel.

– Dave S. Collingridge

Chapter 2

Evolution's Allure of Disbelief

Dave S. Collingridge

This institution [BYU] must turn out the best scientists. We want them to turn out the best philosophers to be found in any of the schools. But we want those scientists and philosophers thus trained to measure the theory of their science and philosophy by the truths of the gospel of Jesus Christ.
–Harold B. Lee

The Atheistic Surge

Over two thousand years ago Korihor taught that there was no God. Because there was no supreme being, he argued, there was no crime, no sin, and no accountability to a higher power. He also taught that everyone fared according to the impartial laws of nature and there was no need for people to work out their salvation before the Lord.

Korihor's teachings apparently appealed to many people living back then. The scriptures tell us that he was successful in leading people away from God. His atheism led many to believe that "when a man was dead, that was the end thereof" (Alma 30:18).

In our time there has been a resurgence of this same atheistic doctrine. It has found favor among those wanting to unburden themselves from a moral responsibility to a higher power. It has also found favor among those who, having put more faith in science than God, willingly deny any involvement of a supreme being in the creation.

Unwittingly Darwinism has contributed to this atheistic surge. Strictly speaking, evolution is a scientific idea about the origins of species and, as such, has nothing to do with religion. However, because evolution is an explanation about the origins of life, it makes assumptions that extend into the religious domain.

An evolutionary assumption that spills over into religion is that all living things, including humans, evolved through blind and undirected processes. In other words, deity was not required to create. Rather than keeping this atheistic belief to themselves, evolutionists are broadcasting it over the internet, in bestselling books, in classrooms, and on television. They want to convert people to their way of thinking.

According to atheists, because Darwinism has given us a completely natural explanation for the origins of species, it is no longer necessary to believe in divine creation. As one atheist put it, "Darwin made it possible to be an intellectually fulfilled atheist."[1] Another atheist said, "If evolution is true, then none of the things that deeply religious people want to be true are in fact true."[2] And another boldly declared, "If humankind evolved by Darwinian natural selection, [then] genetic chance and environmental necessity, not God, made the species."[3] Their message: If you've got Darwin, you don't need God.

Example of Evolution's Allure

Let's take a look at a real-life example of how evolution's atheistic allure weakened one man's faith in God.

In the early 1800s, a young boy grew up in England a firm believer in the teachings of the Bible. At a very early age he was drawn to the study of nature. By the time he reached his early twenties, he had become an accomplished naturalist. He eventually married a devout Christian woman and became a dedicated husband and loving father.

Despite having been born into a Christian home and having married a devout Christian, his faith began to wane as he studied evolution. From his study of evolution, he eventually came to the conclusion that it was no longer necessary to attribute the origins of species to deity. His faith gradually weakened until he completely rejected the role of divinity in the creation. The theory of evolution opened the gates of disbelief, and this man, who was once a devout believer, stepped through.

This man is none other than Charles Darwin.

The following statements are taken from the Life and Letters of Charles Darwin, a collection of autobiographical quotes from Darwin's life.[4]

Darwin was once a firm believer in the Biblical account of the creation.

> I did not then in the least doubt the strict and literal truth of every word in the Bible, I soon persuaded myself that our Creed must be fully accepted. (p. 45)
>
> Whilst on board the Beagle I was quite orthodox, and I remember being heartily laughed at by several of the officers (though themselves orthodox) for quoting the Bible as an unanswer-

> able authority on some point of morality. (p. 307-308)
>
> Formerly I was led by feelings . . . (although I do not think that the religious sentiment was ever strongly developed in me), to the firm conviction of the existence of God, and of the immortality of the soul. In my Journal I wrote that whilst standing in the midst of the grandeur of a Brazilian forest, "it is not possible to give an adequate idea of the higher feelings of wonder, admiration, and devotion, which fill and elevate the mind." I well remember my conviction that there is more in man than the mere breath of his body. (p. 311)

Evolution's allure of disbelief gradually diminished his faith.

> The old argument from design in Nature . . . which formerly seemed to me so conclusive, fails, now that the law of natural selection has been discovered. We can no longer argue that, for instance, the beautiful hinge of a bivalve shell must have been made by an intelligent being, like the hinge of a door by man. There seems to be no more design in the variability of organic beings, and in the action of natural selection, than in the course which the wind blows. (p. 309)
>
> [T]he presence of much suffering agrees well with the view that all organic beings have been developed through variation and natural selection. (p. 311)
>
> [Once] I deserved to be called a Theist. This conclusion was strong in my mind about the time, as

> far as I can remember, when I wrote the 'Origin of Species'; and it is since that time that it has very gradually, with many fluctuations, become weaker. (p. 313).
>
> I gradually came to disbelieve in Christianity as a divine revelation . . . . Thus disbelief crept over me at a very slow rate, but was at last complete. The rate was so slow that I felt no distress. (p. 308-309)

Darwin was once a devout believer in the Lord and the Bible. Evolution's allure of disbelief gradually eroded his faith. The more he studied evolution, the more his faith weakened. This process continued until he stopped believing altogether. Just as people feel no distress when the adversary slowly binds them with a flaxen cord (2 Nephi 26:22), Darwin felt no distress when evolution slowly eroded his faith.

Avoiding Deception

Like Darwin, others are at risk of having their faith diminished by evolution. Perhaps nowhere is this more apparent than in schools where evolution's atheistic allure has been hard at work for generations.

In many high schools, colleges, and universities, evolution is taught as though it has been proven beyond a reasonable doubt. Some students end up thinking that if the best biological scientists say their ancestors were hominids, then it must be so. Not wanting to appear unscholarly or unscientific, some Christian students end up abandoning their religious beliefs and put their trust in Darwinism. Ezra Taft Benson hinted at evolution's allure of disbelief when he said, "Students at universities are sometimes so filled with the doctrines of the world they begin to question the doctrines of the gospel."[5]

Lehi's dream is a good description of the pressures put on Christians to accept evolution. In Lehi's dream, when the mist of darkness arose some people let go of the iron rod, wandered off, and became lost. In much the same way, when surrounded by falsehoods in Darwinism, some believers let go of the word of God, stray from revealed doctrine, and abandon long-held religious views.

There were others in Lehi's dream who, after reaching the Tree of Life, wandered off when they saw people in the great and spacious building ridiculing them. The same sort of thing is happening today. Some believers abandon their faith in inspired teachings and accept Darwinism so they don't have to face the scorn of the scientific community. For these people, being accepted by the scientific community is more important than holding fast to religious beliefs.

If evolution is a potential breeding ground for disbelief, what should be done? First of all, there is no need to discourage children from studying evolution in school. Darwin's ideas contain some truths, such as that natural selection creates microevolutionary variations within a species. It is also important to note that all of evolution does not need to be proven true before it is taught in schools. If all aspects of scientific theories must be proven true before being taught in schools, there would be little left to teach. Like most theories, Darwinism contains a mixture of truths and untruths. This should not prevent it from taking its rightful place in scientific curriculum.

A more reasonable approach to dealing with Darwinism comes from Section 91 of the Doctrine and Covenants. In 1833 while at Kirtland, Ohio, the Prophet Joseph Smith was working on an inspired translation of the Old Testament. When he came to the Apocrypha, a collection of ancient writings in the Protestant Bible,[6] he asked the Lord whether it should be translated along with the rest of the Bible.

The Lord's reply is recorded in Doctrine and Covenants section 91. He replied that in the Apocrypha, "There are many things contained therein that are true" (verse 1), but also "There are many things contained therein that are not true" (verse 2). The Lord did not tell Joseph Smith to avoid the Apocrypha because of the untruths it contained. Rather, He counseled, "whoso readeth it, let him understand, for the Spirit manifesteth truth; and whoso is enlightened by the Spirit shall obtain benefit therefrom" (verses 4-5).

The Lord counseled people who read the Apocrypha to read it with the companionship of the Spirit. The key is studying within the light of the Spirit. With the Spirit as our guide, we can benefit from the truths in the Apocrypha and avoid becoming ensnared by falsehoods.

As with the Apocrypha, evolution should be studied within the light of the Spirit. Having the companionship of the Spirit will enable us to identify truths in evolution and avoid becoming ensnared by falsehoods.

How do we gain the companionship of the Spirit when studying evolution? One way is to listen to the still small voice as we study. The Spirit's guiding influence may especially be felt as we read scriptures on the creation. There are also several gospel-themed books on the creation. Elder Ezra Taft Benson mentioned one such book in a 1970 general conference address. Speaking to parents, he advised, "If your children are taught untruths on evolution in the public schools or even in our Church schools, provide them with a copy of President Joseph Fielding Smith's excellent rebuttal in his book Man, His Origin and Destiny."[7]

The Spirit is key to avoiding the atheistic allure of evolution. It is the bulwark that prevents doctrines of disbelief from eroding our faith. Through humility, obedience to God's commands, and treasuring up the words of the Lord, we invite the Spirit into our hearts and minds. The Spirit will, at times, guide our thoughts about what to believe when we study secular ideas. It can also give us impressions of when something sounds right and when something sounds wrong.

Elder David A. Bednar taught this principle when he said:

> As you appropriately seek for and apply unto the spirit of revelation, I promise you will walk in the light of the Lord. Sometimes the spirit of revelation will operate immediately and intensely, other times subtly and gradually, and often so delicately you may not even consciously recognize it. But regardless of the pattern whereby this blessing is received, the light it provides will illuminate and enlarge your soul, enlighten your understanding, and direct and protect you and your family.[8]

On the Mount of Olives the Savior promised that "whoso treasureth up my word, shall not be deceived" (JST — Matthew 1:37). This promise is as true today as it was 2000 years ago.

Notes

[1]Richard Dawkins, *The Blind Watchmaker: Why the Evidence of Evolution Reveals a Universe Without Design* (New York: W. W. Norton and Company, 1996), 10.

[2]William Provine, in Russell Stannard, *Science and Wonders: Conversations about Science and Belief* (Boston: Faber and Faber, 1998), 45.

[3]Edward O. Wilson, *On Human Nature* (Boston: Harvard University Press, 2004), 1.

[4]*The Life and Letters of Charles Darwin, Including an Autobiographical Chapter* (London: John Murray, 1887). Available at http: //darwin-online.org.uk/content/frameset?itemID=F1452.1&viewtype=text&pageseq=1

[5]Ezra T. Benson, Priesthood Leadership address, April 4, 1986.

[6]Richard L. Bushman, *Joseph Smith: Rough Stone Rolling* (New York: Alfred A. Knopf, 2005), 137.

[7]Ezra Taft Benson, “Strengthening the Family,” *General Conference*, October 1970.

[8]David A. Bednar, ”The Spirit of Revelation,” *General Conference*, April 2011.

Chapter 3

Fact or Theory?

David M. Barker

Facts, as usual, are very stubborn things—overruling all theory. —Jules Verne

The Truth of Things

The average individual's exposure to science seems a mix of news reports, documentaries, and a recollection of science classes from days gone by. It might also include remembering lab work, experiments, and something about testing hypotheses.

Science is often pictured as a high-tech, fact-filled realm. Its beauty, usefulness, and sophistication occasionally fill people with a sense of awe. William Broad and Nicholas Wade observed: "The conventional conception of science exerts a powerful fascination because it is based on a highly attractive set of ideals about how science should work. It can accurately be described as an ideology."[1] In reality, science does not always live up to its idealistic expectations.

Take, for instance, the tendency for people to associate "scientific" with things that have been tested and proven.

Things which have been labelled "scientific" are often considered beyond reproach. In reality, this is not always the case. Although science utilizes high-precision methods and sophisticated terminology, the truth is that some aspects of science are highly speculative and untestable. A good example is the carbon-14 dating system. Although a valuable tool, it relies heavily on assumptions and isn't nearly as reliable and accurate as most people suppose.[2] Carbon-14 and other date-estimation techniques provide much of the material in conflict with religious teachings and especially with scriptural chronology. Why? This and many other questions beg for answers.

People also recognize that science is the study of all things physical, or, as Dr. John Pratt described it, "the study of everything that can be observed, [although] true science makes no claim that it is the study of everything that exists."[3] Unfortunately, many restrict their definition of "science" to what they call "natural" phenomena, and exclude any recognition of God. Mainstream science seems to want it so, and some scientists have waged a war to restrict "science" from open consideration of God and His hand in nature. Such an attitude is contrary to a sound approach to discovering truth.

Referring to theories that were entrenched in the 1960s and 1970s, Hugh Nibley expressed his regret that they have "turned some of our best students away from the gospel." But scientists now recognize that some important aspects of those theories "have turned out to be dead wrong!"[4] To believers, it is truly sad when people lose faith; it is tragic when they do so because of flawed theories. Although some say that resolving the conflicts between science and religion is not necessary for their salvation, the personal salvation of others is certainly in peril—as evidenced by the large number of people who attribute their loss of faith to scientific theories—particularly evolution.

When conflicts appear between the wisdom of men and the revelations of God, it is man that is in error! Man's error is either due to an inaccurate understanding of revelations, of scientific truths, or both. Truth is truth from whatever source. In 1839, Joseph Smith wrote: "Mormonism is truth; and every man who embraces it feels himself at liberty to embrace every truth: consequently the shackles of superstition, bigotry, ignorance, and priestcraft,[5] fall at once from his neck."[6]

There is no incompatibility between truths in science (uncontaminated by false notions) and truths in religion (also uncontaminated). Truths in religion are readily available in scriptures and inspired teachings. However, truths in science are somewhat more difficult to find. The difficulty lies in finding truths amidst the multitude of scientific theories, suppositions, and misinterpretations. "After all," Nibley wrote, "how many wrong answers are there to a problem? As many as you want. But how many right ones? Very few."[7]

In reality, there is only one way a particular event happened, yet empirical descriptions are sometimes vague, complicated, and confusing. Moreover, theories that are supposed to explain events sometimes fail to adequately capture the reasons why something happened.

Students of science study popular theories, regardless of whether those theories are right or wrong. It would be well for them to heed the following advice given to Dr. Henry Eyring[8] by his father. As he was leaving for college, Henry Eyring's father urged him to "Learn everything you can, and whatever is true is a part of the gospel."[9] Eyring's advice implies that we should be cautious about what we accept as scientific truth. This concept was echoed years ago by Albert Einstein who urged people to not "take the present state of our knowledge very seriously. . . . [because] [f]uture developments might show us how to reach a different conclusion from the evidence."[10]

In a discussion about scientific truth, it is important to consider the one who opposes truth, the adversary. Secular science does not recognize such a being, but believers are aware of his plan to disrupt the purposes of God. He has great powers of deception and his aim is clear: "And again, I will give unto you a pattern in all things, that ye may not be deceived; for Satan is abroad in the land, and he goeth forth deceiving the nations" (Doctrine and Covenants 52:14). Neither science nor religion are immune to his buffetings.

Nephi reminded us:

> For behold, at that day shall he rage in the hearts of the children of men, and stir them up to anger against that which is good. And others will he pacify, and lull them away into carnal security . . . and thus the devil cheateth their souls. . . . And behold, others he flattereth away, and telleth them there is no hell; and he saith unto them: I am no devil, for there is none—and thus he whispereth in their ears, until he grasps them with his awful chains. (2 Nephi 28:20–22)

There are also ongoing debates about whether certain scriptural stories are true. Within the Christian community there are those who believe that the Bible is literally true and unerring in every word, while others think it is just a collection of stories. Similarly, within the Church there is a broad range of opinions regarding appropriate interpretation of scripture. Some take a literal approach to the scriptures, while others, due to their belief in science, say that many scriptural stories are allegorical, mere myths and fables.

For those who claim that "hard to believe" scriptural stories are mere myths, questions arise. Was there really one in the pre-mortal existence who said: "We will go down, for there is space there, and we will take of these materials, and we will make an earth whereon these may dwell" (Abraham 3:24)? Was there really an angel of God who "had fallen from heaven; wherefore he became a devil, having sought that which was evil before God" (2 Nephi 2:17)? Was Jesus really resurrected? Many consider these and other scriptural stories to be just "faith-promoting" fables rather than accounts of actual events.

How literally should the scriptures be viewed? The Prophet Joseph Smith taught that "the scriptures say what they mean, and mean what they say"[11]—as far as they are accurately understood and translated correctly. If much of how we view the scriptures depends on translation, then what qualifies as a correct translation? Willamowitz-Moellendorf provides some guidance. He described translation as "a statement in the translator's own words of what he thinks the author had in mind."[12] An accurate translation then is one which closely aligns with what the original author had in mind.

Distinguishing Fact from Theory

The conflict between science and religion is not so much caused by a clash of facts and truths as it is caused by a misunderstanding about the meanings of facts, inferences, and theories. While it is generally accepted that theories are built on facts, it should also be remembered that theories and inferences are intended to organize and help make sense of facts.

Elder John A. Widtsoe[13] acknowledged that "The Church holds that the methods used by science to discover truth are legitimate." But he also cautioned:

> In this wholehearted acceptance of science, the Church makes, as must every sane thinker, two reservations: [1] The facts which are the building blocks of science must be honestly and accurately observed. [2] There must be a distinct segregation of facts and inferences in the utterances of scientific men. Readers of science should always keep this difference in mind. Even well-established inferences should not lose their inferential label.[14]

It is unfortunate that scientific claims are often out of touch with Elder Widtsoe's admonition. Scientists are typically bright, but sometimes descriptions of their findings are so poorly phrased that it is difficult to distinguish between what counts as a fact and what counts as an inference. Words such as "estimated," "assumed," "surmised," "inferred," and "derived from" are often used in scientific reporting. Scientists use these words to convince others that their ideas are valid. Though these words may be essential to scientific writing, they tend to weaken scientific discourse because they introduce imprecision. They make it more difficult for people to distinguish between what counts as a fact and what counts as a theory.

Pratt said that "It is extremely important to distinguish between facts and theories in science. . . . They are not always easy to differentiate, and even scientists forget to do it. And the people who write science textbooks nearly always forget to do it."[15] He also pointed out that in scientific usage, "the word 'fact' has several meanings, which can be very confusing. . . . It can mean either 'observation,' 'theory,' or 'truth.'"[16] Thus comes a startling message: if "fact," "observation," and "truth" are sometimes used synonymously, how can anyone expect to adequately differentiate among them? Careful scientists don't confuse facts, inferences, and theories. The point is, too few are careful enough.

Those who wish to learn what part of a text is fact and what part is theory must do additional research and check an author's sources. Sometimes this involves studying the origin and development of a scientific procedure. For example, in their work on geology, Judson and Kauffman published a discussion on time, including a statement typical of many textbooks: "We now know that the Earth is about 4.6 billion years old."[17] Such a statement implies that not only is there no doubt in the authors' minds that the earth is that old, but that it is in fact that old.

What evidence causes someone to "know" that the earth is 4.6 billion years old? How accurate is that figure, and has the earth remained in its current state for that long? Is it an estimate of something else such as the "age" of certain materials from which the earth was made? To understand the degree of truth behind their statement, one must study the date-estimation techniques used to derive that "age," and learn the theories and assumptions on which those techniques are based.[18] It would be more accurate to say: "The earth is estimated to be about 4.6 billion years old, based on the . . . dating system." The age of 4.6 billion is obviously not based on an actual record of observations over that span of years, but on calculations using facts whose interpretation is heavily reliant on theoretical assumptions.

The powerful role of assumptions in science isn't adequately acknowledged. Pratt was keenly aware of this deficiency when he wrote: "All of science is based on various underlying assumptions. Often these are so deeply rooted that the scientists are unaware of just what they are. If they turn out to be wrong, then the entire edifice built upon them could fall."[19] It is essential to acknowledge one's underlying assumptions and recognize when they may be false. Discovering false assumptions brings us closer to the truth. It "is not a defeat for science, but a liberation."[20]

Consensus Theory

Some theories have gained law-like status within the scientific community. When popular acceptance propels a theory to such a lofty realm, the line between fact and inference is often blurred. When a theory becomes law, not much emphasis is placed on distinguishing between the facts that support a theory, and a theory's inferences. Both are considered equally valid. As one scholar put it,

> A 'law of nature' is one of those concepts that slips through your fingers the more you try to grasp it. The most that can be said about a physical law is that it is a hypothesis that has been confirmed by experiment so many times that it becomes universally accepted. There is nothing natural about it, however: it is a wholly human construct.[21]

Broad and Wade similarly stated:

> When a hypothesis has been confirmed a sufficient number of times, it may take on the character of a law, such as the law of gravity. . . . Laws are valued principles in science because they predict and account for large bodies of facts. They describe important regularities in nature. But they don't necessarily explain the facts they describe.[22]

Pratt offered the following: "You cannot prove any theory to be true. . . . Scientific theories are never supposed to be considered to be absolute truth"[23] He added that scientists should not "forget that they are supposed to be ready to discard the theory objectively and without regret when it fails."[24]

Authors of school texts sometimes neglect to follow Pratt's wise advice. Rens van der Sluijs observed that,

> Too often, standard textbooks and dictionaries do not offer a realistic picture of what is known about a given scientific subject. As they tend to suppress anomalous evidence, the illusion of solid proven fact is then allowed to lull the minds of the critical and the curious and slow down the progress of science.[25]

Overstating truth claims and failing to disclose negative evidence typically arises when writers and editors favor information that confirms their biases. Overstating truth claims gives the impression that a theory has been proven beyond a reasonable doubt when it has not. This is why Dr. Henry Eyring's father encouraged him to "Believe everything scholars can strictly prove and suit yourself about the rest."[26]

Evidence and Proof

What constitutes proof? The answer is not as clear as it may seem. Peter Mere Latham[27] noticed that "People in general have no notion of the sort and amount of evidence often needed to prove the simplest matter of fact."[28] Scientists disagree on what is required to prove something true. In some realms of science, proof is not possible—particularly when dealing with historical events. With few exceptions (e.g., mathematics), proof is a combination of individual perception and scientific observation. As Milton Friedman put it, "Factual evidence can never 'prove' a hypothesis; it can only fail to disprove it."[29]

When is something proven? Nibley answered: "Only when an individual has accumulated in his own consciousness enough observations, impressions, reasonings and feelings to satisfy him personally that it is so." Even still, Nibley cautioned, "The same evidence which convinces one expert may leave another completely unsatisfied."[30] Consider that physicists may even disagree on something as certain as gravity. Newton believed that the actions of the Spirit ultimately caused gravity, but as Nibley pointed out, "Newton's own disciples would not tolerate [t]his position; the very thing he took as proof of God, the force of attraction, they promptly converted into the opposite."[31]

Disputes over scientific truths become even more energized when the debate spills over into the religious domain. Darwin's book, *The Origin of Species*, was an immediate sensation: "The first day the publication went on the market, Oct. 1, 1859, the entire edition of 1500 was sold out. Thus it would seem that the book was received as a best seller."[32] After its publication, "a controversy arose. . . . It was a struggle between the Christian theological conception of man and the conception held by science." At the time there was a perception that "You were either for religion or you were for science," and, according to Huxley,[33] "the battle against the doctrine of inspiration . . . was the crucial engagement in the fight."[34]

It is comforting to know that, one day, all grains of truth about the creation will be sorted out from the chaff of error. The current polarization between religious and scientific perspectives is only temporary, but it may be a while before we find all the answers. Although there are many clues to help us reconcile the differences between Darwinian and Biblical accounts, this conflict will not be fully resolved until the Lord returns.

> In that day when the Lord shall come, he shall reveal all things—Things which have passed, and hidden things *which no man knew*, things of the earth, by which it was made, and the purpose and the end thereof—Things most precious, things that are above, and things that are beneath, things that are in the earth, and upon the earth, and in heaven. (Doctrine and Covenants 101:32–34, emphasis added)

Scientific Dogma

Until the Lord returns and reveals all things, free and open expression of scientific ideas is the best way to move closer toward the truth. Unfortunately, some in the scientific community do not appear to live by this standard.

The 2008 documentary *Expelled: No Intelligence Allowed* features scientists who claimed they were dismissed from academic and scientific positions because they openly advocated Intelligent Design. Intelligent Design promotes the idea that life on earth was created by an intelligent source. Although the institutions from which they were removed gave weakly veiled excuses for the dismissals, it seems rather clear that at least most of these scholars were removed because they dared to challenge evolutionary theory, namely the belief that changes are random, rather than directed.

In *Expelled*, Steve Fuller, Professor of Sociology at the University of Warwick, said: "I'm actually a person of the left, and not even particularly religious. I think of myself as kind of a humanist. And I think it's sending a very bad message to religious people who are interested in science, that in some sense, in order to do science credibly they have to leave their religious beliefs at the door."[35] Dr. Gerald Schroeder similarly observed: "There is academic freedom as long as you are on the correct side."[36]

Other sources suggest that religious scholars are being pressured to go along with the status quo in science. Jeremy Dunning-Davies wrote, "It is well documented that adopting certain stances will result in an inability to publish in the majority of the so-called high impact academic journals."[37] The claims in a 1998 book titled *Silencing Scientists and Scholars in Other Fields* are particularly alarming. The author found evidence of "suppression in academia, science and the art world," and "reveals how far [we've deviated] from [the] ideals of truth and honesty [in] most research."[38]

Pressure to accept the status quo in science was experienced firsthand by Dave Collingridge. In May 2009, Collingridge gave a presentation on science's relationship to deity at a Mormon Scholars in the Humanities conference. Collingridge's main point was that the current scientific culture intentionally excludes God, and that this has not always been the case. Quoting Collingridge, Molly Farmer from the Deseret News reported:

> God has been both passively and actively pushed out of the study of the natural world—and the shift happened despite the fact that some of the world's greatest scientists were also believers. . . . A few movements are responsible for rendering science "Godless and God-hostile." During the 18th century Enlightenment, writers such as Voltaire "downplayed references to God" when they rewrote the treaties of past scientists and made them available to the masses.

> Consequently, many people aren't aware of the way these scientists coupled scientific beliefs with spiritual. . . . One evidence of the movement is the way it turns "freedom of religion into freedom from religion['s influence]." Collingridge said that when he was a student at two different liberal-arts colleges, he was surprised at how all points of view were accepted so long as they weren't religious in nature.[39]

Theories are important to science's success. They represent mankind's best explanations for the facts we find in the world around us. But when science accepts theories as incontestable to the point of excluding legitimate, competing viewpoints, the quest for truth is hindered. As Reid E. Bankhead noted, "Only when theories have been held aloft as unchanging facts or guides to life, have they become dangerous in the search for truth."[40]

Uniformity

In many natural science textbooks, the present is viewed as the key to the past. Many of the natural processes that we see today are the same processes that are believed to have been operating for millions of years. This concept is called uniformity. When applied to evolution, uniformity refers to slow and steady change. Evolutionists claim that life evolved over millions of years and it is still evolving today—it is just too slow to be noticed. Taking note of this uniformity assumption, Nibley wrote:

> We have all grown up in a world nurtured on the comfortable Victorian doctrine of uniformitarianism, the idea that what happens in this world is all just more of the same. . . . [T]he same

> forces that are at work on the earth today were at work in the same manner, with the same intensity and the same effects at all times past and will go on operating inexorably and irresistibly in just the same way forever hereafter. [41]

A problem with evolution's idea of slow and steady change is that it tends to contradict sudden and catastrophic changes that we find mentioned in the Bible. Ever since the early 1800s, uniformitarianism has all but replaced Biblical catastrophic accounts. Donald W. Patten described the change from catastrophic to uniform changes in this way:

> Anti-spiritual humanists, like Voltaire or Kant, usually applauded anything which tended to discredit the Genesis record of catastrophes. Thus the doctrine of uniformitarianism was born and nurtured. . . . Evolution and uniformitarianism practically require agnosticism, and they made atheism increasingly respectable. . . . The viewpoints of the early catastrophists became outmoded and were gradually discarded, and then they were all but forgotten.[42]

Others attribute the sweeping acceptance of uniformity to the influential teachings of Charles Lyell."[43] David Salkeld, who was critical of Lyell's influence, wrote,

> Lyell's theory triumphed over a common-sense approach to the geological evidence and this victory of theory over common-sense was perhaps his most pernicious legacy, because we find it appearing throughout scientific disciplines today. . . . Another of Lyell's legacies is the "million years syndrome": the way in which, where other people had talked of thousands of years,

> he spoke of millions—just like that! He didn't bother to say how he got to millions; he just added three noughts.[44]

According to Victor Clube and Bill Napier, "The uniformitarian view gradually prevailed so that by the middle of the nineteenth century, catastrophism was dead in scientific circles. Of course the advent of Darwinism was apparently the final blow."[45]

Despite uniformity's popularity today, the New Testament seems to cast it in a less than positive light:

> Knowing this first, that there shall come in the last days scoffers, walking after their own lusts, And saying, Where is the promise of his coming? for *since the fathers fell asleep, all things continue as they were from the beginning of the creation.* For this they willingly are ignorant of, that by the word of God the heavens were of old, and the earth standing out of the water and in the water. (2 Peter 3:3–5, emphasis added)

Reid E. Bankhead, President of the Cumorah Mission of The Church of Jesus Christ of Latter-day Saints from 1966 to 1969, suggested to me that these verses refer to uniformity. In reference to these verses, another scholar stated: "Have all things continued as they were? Has history gone on as usual? A great many things have taken place which do not conform to the iron-clad law of uniformity."[46]

Uniformity need not contradict the scriptures—if a less rigid approach is taken. Trevor Palmer commented:

> In 1831, the first Director of the Geological Survey, Henry de la Beche, wrote, "The difference in the two theories is in reality not very great; the question being merely one of intensity of forces,

> so that probably, by uniting the two, we should approximate nearer the truth."
>
> When addressing the Geological Society in 1869, Huxley stated: "To my mind there appears to be no sort of theoretical antagonism between Catastrophism and Uniformitarianism; on the contrary, it is very conceivable that catastrophes may be part and parcel of uniformity."[47]

Unfortunately, the concept of a mix of uniformity and catastrophism was overwhelmed by the strictly uniformitarian worldview. Uniformity has had an extremely strong hold on mainstream science, particularly geology and evolution, for more than a century. Until recently, that grip has been tight enough to exclude most catastrophic events—and all things biblical. Although headway has been made in undoing uniformity dogma, there is still progress to be made.

Conclusion

Science's challenges with discovering truth are compounded by a failure to adequately distinguish between facts, theories, and calculated guesses. For those of us who are interested in finding and keeping truth, it is important to recognize that, far too often, words are used in ways that persuade people to accept unproven scientific ideas.

Weaknesses in popular theories like evolution have convinced me that people should not lose their faith because of "the wisdom of men"—no matter how persuasive the arguments may appear. To prevent people from losing their faith, I've presented three ideas to consider. One is to adequately distinguish between scientific facts and scientific inferences, and guard against inferences that conflict with scripture. Another is to recognize that science cannot fully test some of its popular theories—especially certain aspects of evolutionary theory. And another is to remember that no matter how persuasive a theory may sound, it must not be taken as a disproof of revealed truth.

Notes

[1]William Broad and Nicholas Wade, *Betrayers of the Truth*, (New York: Simon & Schuster, 1982), 15.

[2]For details about some of the weaknesses in the C14 dating system, see *Science and Religion: Reconciling the Conflicts*, by David M. Barker, Chapter 3. A version may also be seen at the author's website http://davidmckaybarker.com/science-and-religion-2/carbon-dating-strengths-and-weaknesses/.

[3]John P. Pratt, "Has Satan Hijacked Science?" *MeridianMagazine.com.*

[4]Hugh Nibley, *Old Testament and Related Studies.* (Salt Lake City: Deseret Book, 1986), 57.

[5]Priestcraft: "Priestcrafts are that men preach and set themselves up for a light unto the world, that they may get gain and praise of the world" (2 Nephi 26:29).

[6]*Teachings of Presidents of the Church: Joseph Smith.* (Salt Lake City: The Church of Jesus Christ of Latter-day Saints, 2007), 264.

[7]Hugh Nibley, *Temple and Cosmos.* (Salt Lake City: Deseret Book, 1992), 504.

[8]Dr. Henry Eyring, father of President Henry B. Eyring, of the First Presidency of the Church.

[9]Henry Eyring, *Reflections of a Scientist.* (Salt Lake City: Deseret Book, 1983), 1.

[10]Charles H. Hapgood, *Earth's Shifting Crust: A Scientific Key to Many of Earth's Mysteries.* (New York: Pantheon Books, 1958), 364–365.

[11]Joseph Smith, *Teachings of the Prophet Joseph Smith*, ed. Joseph Fielding Smith (Salt Lake City: Desert Book Company, 1976), 264.

[12]Hugh Nibley, *Mormonism and Early Christianity.* (Salt Lake City: Deseret Book, 1987), 249.

[13]Elder Widtsoe was a scientist, as well as a member of the Quorum of the Twelve Apostles from 1921 until his death in 1952.

[14]John A. Widtsoe, *Evidences and Reconciliations.* (Salt Lake City: Bookcraft, 1943), 127–128.

[15]John P. Pratt, "Fact or Theory?" http://www.johnpratt.com/items/astronomy/science.html.

[16]Pratt, "Fact or Theory?"

[17]Sheldon Judson and Marvin E. Kauffman. *Physical Geology*, 8th ed. (Englewood Cliffs, New Jersey: Prentice Hall, 1990), 3.

[18]For some summary details, see chapter 5 of *Science and Religion: Reconciling the Conflicts* by David M. Barker.

[19]John P. Pratt, "Strengths and Weaknesses of Science." *Meridian Magazine.* (28 Dec 2000), 7.

[20]Phillip Johnson, *Darwin on Trial.* (Downers Grove, Ill: InterVarsity Press, 1991), 154.

[21]Editorial, "Breaking the Laws." *NewScientist.* (April 29, 2006), 5.

[22]Broad and Wade, *Betrayers of the Truth.* 16.

[23]Pratt, "Fact or Theory?" 2.

[24]Pratt, "Strengths and Weaknesses of Science." 6.

[25]Van der Sluijs, Rens, "The Unwavering Truth." *Chronology and Catastrophism (C&C) Workshop.* The Society for Interdisciplinary Studies (SIS), (2010:1), 40.

[26]Henry Eyring, *Faith of a Scientist.* (Salt Lake City: Bookcraft, 1967), 183.

[27]Peter Mere Latham (1789–1875) was considered a great medical educator.

[28]John Bartlett, *Bartlett's Familiar Quotations*, 15th Ed, (Boston: Little, Brown and Company, 1980), 464.

[29]Bartlett, 880.

[30]Nibley, *Since Cumorah.* viii.

[31]Hugh Nibley, *The Ancient State.* (Salt Lake City: Deseret Book, 1991), 409–410.

[32]Donald Wesley Patten, *Biblical Flood and Ice Epoch: A Study in Scientific History*, (Seattle: Pacific Meridian Publishing Co. 1966), 12.

[33]Huxley, Thomas Henry (1825-1895), "British biologist, best known for his active support of Charles Darwin's theory of evolution." *Encarta Encyclopedia.* 2004.

[34]Nibley, *The Ancient State*, 403.

[35]"Expelled: No Intelligence Allowed." Premise Media Corp. DVD. 2008, 58:03.

[36]"Expelled," 1:25:53.

[37]Jeremy Dunning-Davies, "Science in Turmoil—Are we Funding Fraud?" http://www.thunderbolts.info/thunderblogs/guest_jdd.htm.

[38]Jill Abery, Review of *Silencing Scientists*, by G. Moran. *Chronology and Catastrophism (C&C) Review.* The Society for Interdisciplinary Studies (SIS), (2001:1), 50.

[39]Molly Farmer, "Science Should Include God." *Deseret News.* (May 14, 2009).

[40]Reid E. Bankhead, *Fall of Adam, The Atonement of Christ, and Organic Evolution.* (Pleasant Grove, Utah: National Research Group). (undated), 12.

[41]Nibley, *Temple and Cosmos.* 451.

[42]Patten, *Biblical Flood and Ice Epoch.* 2.

[43]Roy R. Lemon, *Principles of Stratigraphy,* (Columbus, Ohio: Merrill Publishing Co. 1990), 30. Charles Lyell (14 November 1797 – 22 February 1875) "was a British lawyer the foremost geologist of his day. He is best known as the author of Principles of Geology, which popularized . . . uniformitarianism." http://en.wikipedia.org/wiki/Charles_Lyell.

[44]David Salkeld, "Genesis and the Origin of Species." *C&C Review.* SIS, (2002:1), 10.

[45]Victor Clube and Bill Napier, *The Cosmic Serpent: A Catastrophist View of Earth History.* (New York: Universe Books, 1982), 13, 94.

[46]Elmer G. Homrighausen, "The Second Epistle of Peter." In, *The Interpreter's Bible.* Vol. 12, (New York: Abingdon Press, 1957), 198.

[47]Trevor Palmer, "Uniformitarianism, Catastrophism and Evolution." *C&C Review.* SIS, 1996:1, 7, 13.

Chapter 4

Evolutionary Evidences

Dave S. Collingridge

They who have presumed to dogmatize on nature, as on some well investigated subject, either from self-conceit or arrogance, and in the professorial style, have inflicted the greatest injury on philosophy and learning. For they have tended to stifle and interrupt inquiry exactly in proportion as they have prevailed in bringing others to their opinion.
—Francis Bacon

Darwin's Telescope

In the early 1600s, Italian scientist Galileo Galilei received word that a Dutch lens grinder had made a revolutionary discovery in optics. The Dutchman discovered that when certain lenses were combined, distant objects were made to appear closer. Upon hearing this news, Galileo applied himself to the study of refraction and attempted to replicate the Dutchman's discovery.

Galileo made his own telescope by preparing a lead tube fitted with a concave lens at one end and a convex lens at the other. So effective was his telescope design that he built more and sold them to interested persons. While others were mainly interested in using the telescope for viewing objects on land and sea, Galileo turned his telescope toward the heavens.

For more than 2000 years, objects in the heavens were thought to be perfect and unalterable. Contrary to expectations of the moon being smooth, Galileo observed a rough lunar surface covered with cavities and jagged peaks. People believed that the sun was a flawless sphere. Galileo discovered that the sun has dark, irregular blemishes (sunspots) which move across its surface. And contrary to the commonly held belief that everything in the heavens revolved around the earth, Galileo observed small stars (moons) moving around Jupiter. He also observed Venus passing through moon-like phases, something that would be expected to happen if Venus moved around the sun.

Because Galileo's observations with his telescope contradicted long-held philosophical and religious beliefs about the nature of the heavens, his discoveries were not immediately accepted. Some people argued that the things he saw were mere illusions created by his telescope. Others said the rough lunar surface he saw in his telescope must be covered by a smooth, transparent material, thus preserving the traditional view of the moon being a perfect, heavenly sphere. Even the leading philosopher in Pisa, Italy refused to look through Galileo's telescope for fear of what he might find.

Evolutionists certainly feel their theory is facing the same kind of resistance from religious folks that Galileo faced long ago. Some evolutionists think that if religious people would just look at the evidence for evolution, they would realize Darwin's ideas are true. They have good reason to be optimistic; many believers have hopped on the evolutionary bandwagon after seeing the evidence for evolution.

Are the rest of us who continue to reject evolution like the 17th century Italians who would not accept Galileo's observations? Are we like the Pisa philosopher who would not look through Galileo's telescope for fear of what he might find? Perhaps some of us are apprehensive about taking a close look at evolution's evidence for fear of finding compelling evidence that contradicts our cherished religious beliefs.

Let's take a look at recent evolutionary science. Let's peer through "Darwin's telescope," so to speak, and examine the evidence. A review of the facts will show that the current body of evidence does not lead to the inevitable conclusion that evolution has been proven beyond a reasonable doubt.

Biological Similarities

Biological similarity between species is sometimes mentioned as proof of evolution. Different species resemble each other anatomically (e.g., two eyes, two forward limbs, one head, one heart). Similarities are especially evident during prenatal development. Some animal embryos (e.g., chicken, dog, and fish) look surprisingly similar to human embryos. There are also genetic similarities between humans and animals. According to latest estimates, humans and apes are about 95% similar, genetically speaking.

Evolutionists explain these similarities by saying we share a common ancestor. According to evolution, millions of years ago different environmental pressures (i.e., natural selection) acted on random mutations in ways that produced different branches of evolutionary history. Although the branches ended up producing different species, the species retained the genetic and anatomical remnants of their shared ancestors.

Evidently there is another explanation for biological similarities. Biological similarities exist among life forms because all life was created by the same designer. The Lord apparently saw no need to completely reinvent anatomical structures and their basic functions each time He created a new species. So the answer to why species have anatomical similarities is that they share a common designer, not a common ancestor.

To clarify this point, let's consider an analogous example from fine art.

In 2014 in Toulouse, France, homeowners went into their attic to repair a leaky roof. In the attic they discovered a painting that had been stored there by a previous owner. The painting depicts a scene from chapter 13 of the Book of Judith in the Apocrypha. That chapter describes a scene where Judith, in an effort to free her people from the tyranny of the Assyrian ruler, entered the enemy's camp and cut off the head of the Assyrian King.

There is no visible signature on the painting identifying the artist. The true identity of the painting's artist remained a mystery until experts began studying the painting's characteristics. Experts saw similarities between the painting's artistic style and the artistic style in other paintings. By identifying similarities between the Judith painting and other paintings, experts concluded that the creator was an early 1600's artist named Caravaggio. If the painting was not created by Caravaggio, it was created by someone, perhaps a student, who emulated Caravaggio's style.

Similarities in artistic style led experts to conclude that the Judith painting and other paintings share a common designer. In much the same way, similarities in the biological structure of animals suggest they were created by a common Creator. (As a side note, the painting's value was estimated at over 100 million euros. That's enough to make others check their attics!)

Imperfections

Evolutionists who vigorously oppose divine creation like to point out that the Creator is a poor designer. They say that if we had been created by an omnipotent deity, our bodies would have little or no imperfections. In their minds, if there was an all-powerful Creator, He would have used His wisdom and power to design a better body, a body that is resistant to the effects of disease and aging.

Evolutionists also claim that an all-powerful Creator would not have designed vestigial (useless) structures and "junk" (useless) DNA into our bodies. The human tail bone is an oft-cited example of a vestigial structure. If the tail bone serves no purpose to humans, why would an intelligent being outfit us with one? It is illogical for humans to have tail bones, they say, because we don't have tails. Of course, evolutionists have a logical explanation for the tail bone. They claim it is a structure left over from when our primate ancestors used tails to swing through trees.

Saying the Lord is a poor designer because He made imperfect bodies reflects a misunderstanding of His purposes. There are divine reasons why the human body looks and functions the way it does. Isaiah 55:8 reminds us that His thoughts are not our thoughts, and His ways are not our ways. It is prideful to assume God should have created the human body in a manner consistent with human ideals about how it should look and function.

Most people can probably think of an extra ability that they would like to have, such as flying or breathing under water. I am sure there were times when my mother would have liked an extra pair of eyes on the back of her head. Rear-facing eyes would have helped her keep watch over me and my 10 siblings. Contrary to what mankind thinks is the best placement for eyes, eyes in the back of the head was not the Lord's way of doing things, much to my relief while growing up.

Claiming that our bodies would function more effectively if we had been created by an omnipotent being also reflects a lack of understanding about the Plan of Salvation. We were placed on an imperfect world with imperfect bodies to grow and prove ourselves worthy of exaltation. The toils of mortal existence are a proving ground for eternal life. Mortal imperfections keep us humble and on our knees like few other things can.

In 2 Corinthians 12:7-10 we read that the Apostle Paul suffered physical afflictions. Despite asking the Lord to remove his afflictions, they continued (verse 8). The scriptures suggest that Paul's physical afflictions persisted because of his involvement with several miraculous events. Afflictions prevented him from becoming conceited from all the wonderful things he had seen and done (verse 7).

Thus we see that Paul's afflictions served a higher purpose. They brought him close to the Lord. When his suffering became unbearable, he was strengthened by the power of Christ. So wonderful were the blessings of being strengthened by the Lord that Paul eventually took pleasure in his infirmities (verse 10). He learned that passing through affliction made him stronger. For Paul, the blessing of being strengthened by the Lord far outweighed any afflictions he had to endure.

The idea of physical infirmities making us stronger is counterintuitive from a Darwinian standpoint. From an evolutionary standpoint, physical infirmities make us weaker and reduce our chances of survival. From a spiritual standpoint, physical infirmities make us stronger and bring us closer to the Lord. These represent two very different ways of viewing physical suffering.

The more perfect bodies that evolutionists suppose we should have, if we were created by an omnipotent being, will eventually come. We will receive perfect bodies in the resurrection. All who have ever lived on the earth will be resurrected with immortal bodies that are free from the physical infirmities and sufferings we experience in mortality.

Fossil Record

Evolutionists point to the fossil record in geological strata (ancient layers of dirt and rock) as evidence of evolution. Fossils from less complex species are typically found in deep sedimentary layers that formed long ago. On the other hand, fossils from more complex species are usually found in sedimentary layers nearer to the earth's surface. This fossilization pattern suggests that simpler species predated more complex species.

Darwinism explains this fossilization pattern by saying that life gradually evolved from less complex species into more complex species. This change is commonly believed to have happen through a couple of processes. One process is natural selection. Natural selection refers to environmental circumstances that favor the survival of species with certain traits. The other process is random variation. Survivability traits typically arise from random changes in the nucleotide sequence of DNA, or from random changes in the arrangement of chromosomes.

If the simple-to-complex fossil pattern was not created by evolutionary processes, how did it come about? A possible explanation is that the fossilization pattern resulted from reoccurring floods and other catastrophic geological events, events used by the Lord to create freshwater lakes, rivers, and streams. Catastrophic events may have served another purpose — to prepare the earth's soil for sustaining plant life. As any good gardener knows, soil must contain microbes and decomposing vegetation for plants to grow. Flooding and other earth-disrupting events could have been used by the Lord to mix vegetation with fine silt, thus creating topsoil capable of sustaining life on earth.

The simple-to-complex fossilization pattern may have also been influenced by the order in which the Lord brought life forms of different complexity to the earth. Simple life forms like plants were brought to the earth first. When that stage of the creation was complete, more complex life forms were brought to the earth, including all manner of insect and marine life. These creatures were then followed by more complex creatures such as birds and land animals.

When we combine catastrophic events and creatures being brought to the earth at different times, a possible explanation for the simple-to-complex fossil pattern emerges. Catastrophic events which occurred at different time points during the creation could have fossilized different life forms. Floods that occurred earlier in the creation would have fossilized plant life, but not animal life because animals had not yet been placed on the earth. As increasingly complex life forms were brought to the earth, catastrophic events would have fossilized these life forms in newer sediment, above the sediment containing less complex species.

This sounds like a reasonable explanation, but because no one was around when these events happened, I must resort to educated guessing, as do the evolutionists. The Lord knows why fossils are layered from the simple to complex. He was there when it happened. We were not. We will most likely learn what caused the fossilization pattern when the Savior returns and reveals secrets hidden in the earth (Doctrine and Covenants 101:33).

California Salamander

The Californian salamander is sometimes mentioned as an example of evolutionary change. Long ago, a species of salamander in northern California split into two groups as it migrated south around the dry Central Valley. One group took the easterly route around the Central Valley. The other group took the westerly route around the Central Valley.

Because the easterly route is largely forested and the westerly route is largely coastal, the two groups of salamanders were subjected to different environmental pressures. Different environmental pressures on both sides of the valley led to different physical characteristics. Eventually, when the migratory paths of the two groups came into contact at the southern tip of the valley, they did not interbreed. The two groups were reproductively isolated from one another.

Evolutionists are quick to point out that reproductively isolated creatures are different species. The fact that the two groups of salamanders would not interbreed is considered proof that they evolved into different species.

Reproductive isolation in the California salamander is an example of microevolutionary change. Microevolution refers to subtle variations within life forms. While change in the appearance of the two groups of salamanders may have prevented them from reproducing with one another, they were still salamanders when they came into contact at the southern tip of the Central Valley. Evolution did not change the salamanders into lizards or frogs.

Peppered Moth

We sometimes hear that evolution has been subject to much testing, and passed. We also hear that because of its overwhelming success as a theory, Darwinism has been proven beyond a reasonable doubt. Supposedly it is as certain as

Gravity and Relativity. To support these claims, evolutionists cite instances where natural selection created variation within a species. A favorite among evolutionists is the peppered moth.

A couple of hundred years ago, England's peppered moth was largely colored white with black specks. When the moth rested on trees with light-colored bark, its white color and black specks blended in with the tree bark. The moth's color acted like a camouflage, protecting it from hungry birds.

When the Industrial Revolution came along and factories rose up in cities like Manchester, England, soot from coal-fired factories began to darken the white tree bark in heavily industrialized regions. Consequently white peppered moths that landed on dark tree trunks became more visible to birds and were more likely to be eaten.

As the population of white-colored moths decreased, the population of dark-colored peppered moths increased. Environmental pressures brought on by the Industrial Revolution favored survival in dark peppered moths. Dark-colored moths that blended in with soot-covered tree trunks were less likely to be eaten by birds searching for a meal. Thus, dark-colored moths were more likely to survive and produce offspring than white-colored moths. The result is the peppered moth predominantly changed from white to dark.

Color change in the peppered moth is an example of microevolutionary change, not macroevolutionary change. The moth did not change into a different life form. It did not evolve into an organism far safer from preying birds, like a june bug. But because natural selection has been shown to create microevolutionary change in creatures like moths, evolutionists expect us to accept all of Darwinism. In a way, they expect us to accept that living things evolved from one form into another simply because they've demonstrated that evolution produces minor changes within a species.

The Wizard of Darwinism

Seeing as they lack direct evidence for change across life forms, evolutionists sometimes resort to misdirection to convince us that macroevolution is real. They encourage us to pay attention to the evidence for microevolution, while ignoring the absence of direct evidence for macroevolution.

This sort of misdirection resembles a scene from the Wizard of Oz. When Dorothy and her friends enter the Wizard's lair, they are in awe of the Wizard's greatness and power. They are astonished by the Wizard's impressive display of sound, fire, and smoke. When Dorothy's dog pulls back the curtain, the Wizard's deception is exposed. He is speaking into a microphone while busily operating levers and buttons on a control panel. Dorothy and her companions learn he is not great and powerful. His cover blown, the Wizard makes a last ditch effort to hide his deception. He urges Dorothy and her friends to "Pay no attention to that man behind the curtain!"

In a similar manner, evolutionists fascinate the public with all sorts of arguments and evidence in favor of evolution. Some of the evidence is apparently indisputable, such as that natural selection caused the peppered moth to change from white to black, or that natural selection created two groups of reproductively isolated salamanders. Yet when the curtain of deception is pulled back, we expose evolution's ruse. There are no examples of organisms evolving into different, viable life forms. Not one.

In an attempt to save face, evolutionists typically redirect our attention back to the evidence for microevolution. They expect us to accept that macroevolution is just microevolutionary change over time. When we pull back the curtain on the evidence for major morphological change across life forms, in Oz-like fashion they urge us to "Pay no attention to the absence of direct evidence for macroevolutionary change."

Directly observing major morphological change is the holy grail of evolutionary research. Darwinists know this, which is why they have carried out studies attempting to demonstrate major morphological change. In their minds, if they can show that life is capable of evolving into a different form, they will have solid evidence that life evolved from a common ancestor.

One way evolutionists have attempted to demonstrate major morphological change is by studying organisms that reproduce rapidly. Two organisms that have garnered considerable attention are fruit flies and bacteria.

Fruit Flies and Bacteria

According to evolutionists, the main reason we have not observed macroevolutionary change is that it takes place over long periods of time. Supposedly macroevolution occurs so slowly that we do not notice it taking place.

What if we study species that reproduce at a rapid rate? Assuming macroevolution is real, if we take a species that reproduces at a rapid rate and subject it to environmental pressures, we should see major morphological changes.

Casey Luskin, research coordinator for the Center for Science and Culture at the Discovery Institute in Seattle, Washington, investigated this issue. In 2013, Casey thoroughly reviewed research articles allegedly providing the best evidence for major physical change.[1] The papers he reviewed were cited on a pro-evolution website called TalkOrigins. His findings were revealing.

Luskin looked at studies involving fruit flies, a favorite among evolutionists for their rapid rates of reproduction. In one study, researchers were able to increase the number of hairs on the backs of fruit flies by introducing reproductive pressures into their environment.[2] However, increasing the number of hairs on the backs of fruit flies is an example of microevolutionary change. It is not a demonstration of macroevolutionary change. The hairy-backed fruit flies were still fruit flies. Increasing the number of hairs on the back of a fruit fly no more comprises a new life form than new hair growth on the back of a middle-aged man comprises a new life form.

Casey also looked at studies involving bacteria. Bacteria are capable of producing at an alarming rate. Their ability to reproduce multiple generations over a short period of time makes them good candidates for demonstrating macroevolutionary change.

Luskin cited a study where survival pressures were artificially created in a colony of bacteria through the introduction of a ciliate predator.[3] Ciliates are single-celled organisms that feed on bacteria in aquatic ecosystems. Because larger bacteria in the colony were better able to evade the ciliate predator than smaller bacteria, they were more likely to survive and produce offspring than smaller bacteria.

As a result of introducing a predator into the colony, the size of the bacteria increased over time. Nevertheless, major morphological changes were not observed. Despite increasing in size, the bacteria were still bacteria.

After reviewing several studies on bacterial evolution, Luskin concluded that none demonstrated major morphological change. He is not alone in this conclusion. Repeated failure to produce significant physical changes in bacteria led one prominent British bacteriologist to conclude: "Throughout 150 years of the science of bacteriology, there is no evidence that one species of bacteria has changed into another."[4]

In all the evolutionary studies that Luskin reviewed, not one showed significant morphological change.[5] Although evolution is supposedly happening all around us, direct evidence for change across life forms remains elusive. As one expert put it, "we are in a situation today similar to that experienced by Darwin more than a century ago [wherein] differentiation of species is inferred from copious indirect evidence, but has not actually been observed."[6]

Conclusion

The idea that macroevolution has been proven beyond a reasonable doubt exists only in the minds of those who accept it as such. For skeptics, significant change across life forms has not been proven because it has never been demonstrated.

We've looked through "Darwin's telescope" and found the evidence for macroevolution lacking. The absence of direct evidence for major morphological change means people are justified in rejecting the idea that humans descended from lower life forms. It is perfectly reasonable to reject evolution as an explanation for the origins of species.

Notes

[1]Casey Luskin, *Specious Speciation: The Myth of Observed Large-Scale Evolutionary Change: A Response to TalkOrigins' Observed Instances of Speciation*, http://www.discovery.org/f/8411.

[2]J.M. Thoday and J.B. Gibson, "Isolation by Disruptive Selection," *Nature*, Vol. 193:1164-1166 (March 24, 1962).

[3]Shuichi Shikano, Leo S. Luckinbill, Yasushi Kurihara, "Changes of Traits in a Bacterial Population Associated with Protozoal Predation," *Microbial Ecology*, Vol. 20 (1): 75-84 (July - August, 1990).

[4]Alan Linton, "Scant Search for the Maker," *Times Higher Education Supplement*, April 20, 2001, 29.

[5]Casey Luskin, *Specious Speciation: The Myth of Observed Large-Scale Evolutionary Change: A Response to TalkOrigins' Observed Instances of Speciation*, http://www.discovery.org/f/8411, p. 37.

[6]Theodosius Dobzhansky, and Olga Pavlovsky, "Experimentally Created Incipient Species of Drosophilia," *Nature*, 230 (1971), 289–292.

Chapter 5

Should Latter-day Saints Accept all of Evolution?

David M. Barker

Evolution is a flexible word. It can be used by one person to mean something as simple as change over time, or by another person to mean the descent of all life forms from a common ancestor. . . . In its . . . biological sense, however, evolution means a process whereby life arose from non-living matter and subsequently developed entirely by natural means. That is the sense that Darwin gave to the word. —Michael J. Behe

By the end of my freshman year of college, I had just about become convinced that the Lord used evolution to create life on Earth. However, shortly after arriving in the mission field (Cumorah, 1967–1969), I learned that my Mission President (Reid E. Bankhead, an extraordinary scholar of the scriptures) strongly opposed such a notion. He supported his position with numerous scriptural references. Particularly impressive was a series of scriptural passages and reasoning he called "Revealed vs. Human

Knowledge."[1] The power of those scriptures and the ideas he shared brought me to a realization that the theories of men do not, and should not, take precedence over the scriptures.

President Bankhead also emphasized that the Fall of Adam is fundamental to an understanding of the gospel of Jesus Christ.[2] Adam's Fall from a paradisiacal state has important implications relative to modern science, and particularly to the theory of evolution. Since my mission, I have studied these and related subjects and have found an abundance of information supporting President Bankhead's perspective. I've concluded that his gospel perspective regarding the creation is on a much surer footing than so much of what is taught in schools, textbooks, and television programs (which usually present evolution as though it is a solid and proven fact).

What Part of Evolution Is Fact?

In any discussion on evolution, it is important to be clear about what is meant when one uses the word. Sometimes the word "evolution" simply refers to change. At other times it refers to theories on the origins of species. Failure to be clear about what is meant when referring to evolution can create problems. As Elder John A. Widtsoe[3] pointed out,

> Among people generally, as well as by a group of scientists who should know better, the word [evolution] is used with unpardonable looseness. Especially should the difference between the law of evolution and the theory or theories of evolution be stressed whenever the word is used.
>
> In its widest meaning evolution refers to the unceasing changes within our universe. Nothing is static; all things change. Stars explode in space; mountains rise and are worn down. .

> . . The law of change, an undeniable fact of human experience, is the essence of the law of evolution.[4]

When "evolution" is used to mean "change," there are no disputes. However, when it is used to refer to the origins of species, there is disagreement. Take, for example, the distinction between micro and macroevolution. Microevolution refers to the small-scale changes which are known to occur. Macroevolution, on the other hand, refers to major morphological changes which have not been shown to occur.

Many people remain skeptical about whether macroevolution can happen while others claim that it has happened. As David Salkeld pointed out, the process of macroevolutionary change is popularly believed to be just "microevolution with unlimited time."[5] Thus microevolutionary change within a species is often cited as "proof" of macroevolutionary change.

A symposium was recently held in which an LDS scientist tried to dismiss the distinction between microevolution and macroevolution. Criticizing those of us who maintain that microevolutionary changes are not sufficient evidence for macroevolutionary change, the scientist said, "they believe in inches, but not in feet." This clever phrase caught my attention, but I soon dismissed the slogan as an inappropriate comparison. Inches and feet are merely units of measurement within the same system—there being no issue as to the ability to see and use either or both. However, although small-scale changes (microevolution) have been observed in living things from one generation to another, large-scale changes (macroevolution) have not. To me, this LDS scientist's statement is an example of poor logic and possibly an attempt to skirt the challenges evolutionists face in preaching macroevolution with all its ramifications.

Is there observable evidence showing that macroevolution (change from one type or kind to another) actually happens? Biologist Caroline Crocker challenged the Darwinian view that macroevolutionary events have been proven. She said:

> The development of antibiotic resistance in bacteria is often cited as evidence for evolution. But this is, at best, microevolution. It would be invalid to extrapolate the results to claim they prove macroevolution. . . . [A] species of bacteria called E. coli has undergone 150 years of intensive culture and experimentation, including being genetically engineered. Since bacteria grow very quickly, multiplying from one to millions in a day, it could be expected that evolution into a different form would be evident. Dr. R. Lenski grew this bacterium for 20,000 generations and did not find even one new molecular machine. . . . E. coli is still E. coli and not another species of bacteria, much less something besides a bacterium.[6]

Elsewhere Crocker accused mainstream science of "disguising an atheistic view of life in the garb of science," and claimed that "no one has ever seen a dog turn into a cat in a laboratory."[7] Speaking at a conference of biologists in 1980, Norman Macbeth addressed the problem of the lack of evidence for macroevolutionary change.

> The evolutionists had long neglected this problem preferring to concentrate on changes in coat-color and bristle-number rather than on how a tiny shrew could evolve into a whale or a reptile into a bird. They were convinced that such evolution must have occurred, but they knew of

> no mechanism by which to explain how it was done.[8]

As recently as 1999, a National Academy of Sciences (NAS) booklet listed Darwin's study of "13 species of finches" on the Galapagos Islands as "evidence supporting biological evolution."[9] The NAS booklet also mentions that "more than a thousand species of snails and other land mollusks . . . [are] found only in Hawaii." But aren't all thirteen of Darwin's "13 species of finches" still finches? Aren't finches still birds? And aren't different varieties of snails in Hawaii still snails? The NAS neglected to mention that these variations involved micro not macroevolution changes.

Spontaneous Change—From Disorder to Order?

As mentioned previously, it is true that things change; it seems that nothing in mortality remains constant for very long. But do things spontaneously change from simple to complex as evolution suggests?

When I showed a simple stone arrowhead to a group of students and asked them to identify it, they all seemed to know it was an arrowhead. When I told them that it evolved into that shape, the response was humorous disbelief. If it isn't reasonable to assume that something as simple as an arrowhead can form without deliberate effort, how reasonable is it to assume that all life formed from unguided processes?

In the book *Physical Science Foundations*, Juliana Boerio-Goates makes the following observation: "A diver never dives upward from a swimming pool, just as scrambled eggs never spontaneously unscramble and a pet dog doesn't get younger. Our experience in everyday life shows us that nature tends to run in one direction."[10] That direction is toward decay and disorder.

It may be argued that the notion that living things spontaneously evolve from the simple to complex contradicts the Second Law of Thermodynamics. According to this law, matter naturally moves toward a disordered state. How can nature be inclined to move toward a more disordered state while perpetually evolving from the simple to more complex? Commenting on this conflict, Hugh Nibley said,

> What many are pointing out today is that the mechanistic-evolutionary theory reverses both the direction of time and the order of nature. By the laws of thermodynamics, "left to itself, everything tends to become more and more disorderly, until the final and natural state of things is a completely random distribution of matter. . . ." [Yet] in the words of P. T. Matthews, "The sorting process—the creation of order out of chaos—against the natural flow of physical events is something which is essential to life."[11]

If matter tends to move toward a more disordered state, then how likely is it that complex organisms evolved by chance? Nibley provided more insight.

> When the biologist said that life was wildly improbable, a rare unreasonable event, who would have guessed how improbable it really was? "A human being," writes Matthews, "is at very best, an assembly of chemicals constructed and maintained in a state of fantastically complicated organization of quite unimaginable improbability." So improbable that you can't even imagine it. So "wildly improbable" that even to mention it is ridiculous. So we have no business being here. That is not the natural order of things. . . . So the physical scientists and

> the naturalists agree that if nature has anything to say about it, we wouldn't be here. This is the paradox of which Professor Wald of Harvard says, "The spontaneous generation of a living organism is impossible. . . ." The chances of our being here are not even to be thought of, yet here we are.[12]

Unless acted upon, the natural state of matter is one of disorder and decay. A new car can be purchased, and even if it is kept in a climate-controlled garage, parts will decay. If exposed to natural processes, rust and corrosion ultimately claim it.

Opposition Toward Religious Explanations

It seems that in order to win converts to their position, some advocates of Darwin's theory have set their sights on religion. Nibley provided some insights on anti-religious motives.

> Students commonly assume that it was the gradual amassing of evidence that in time constrained such men to part company with the Bible. Exactly the opposite is the case: long before they had the evidence, they brought to their researches such an unshakable determination to discredit the book of Genesis that the discovery of the evidence was a foregone conclusion.[13]

Nibley elaborated on this hostility toward the Bible.

> It was Darwin's bosom friend and spokesman who blurted out the real issue with characteristic bluntness: "Darwin himself avoided attacking the Bible, but for Huxley . . ." writes J.

> C. Greene, "the battle against the doctrine of inspiration . . . was the crucial engagement in the fight for evolution and for freedom of scientific inquiry." The battle was against revelation, and evolution was the weapon forged for the conflict.[14]

Some of this hostility toward Biblical accounts may stem from evolution's theoretical assumptions. The theory of evolution is a sophisticated structure built from facts, assumptions, inferences, and conjectures. Jonathan Wells identified Darwinism's underlying assumptions when he described it as being "driven by a . . . commitment to materialistic, antitheistic philosophy."[15]

Materialism demands that natural phenomena be explained by natural, not supernatural processes. This antitheistic position is evident in the way some have openly attacked religion. These attacks have heightened the struggle between religious and evolutionary worldviews. Karen Armstrong described this struggle as follows:

> The clash between those who adhere to the scientific theory of evolution and those who believe that the biblical story of the six-day creation is literally true is a struggle between two religions. . . . [S]cientists have offered up an alternative vision of the nature of reality, and those among them who are most opposed to religion can proselytize with as much zeal as an evangelical Christian. For Richard Dawkins, contemplation of the natural world through the eyes of science is a religious experience. . . . Yet Dawkins regards faith as one of the world's great evils, "comparable to the smallpox virus but harder to eradicate".[16]

In a startling interview shown in the movie *Expelled: No Intelligence Allowed*, biologist P. Z. Meyers, who runs a pro-Darwin, anti-religion blog, said:

> Religion is an idea that gives some people comfort, and we don't want to take it away from them. It's like knitting, people like to knit. . . . We're not going to take away their churches. But what we have to do is get it to a place where religion is treated at the level it should be treated. That is, something fun that people get together and do on the weekend. And really doesn't affect their life as much as it has so far. Greater science literacy . . . is going to lead to the erosion of religion, and we'll get this positive feedback mechanism. Whereas religion slowly fades away, we get more and more science to replace it.[17]

When Dr. Peter Atkins, Professor of Chemistry at Oxford University, was interviewed for *Expelled* he said: "Religion . . . is just fantasy basically. It's completely empty of any explanatory common sense, and it's evil as well."[18] C. D. Darlington suggested that the evil stems from the way that religious superstition has "enslaved mankind." He went on to suggest that evolution has "released thinking men from the spell of [religious] superstition."[19]

Never sparing in his criticism of religion, Richard Dawkins claimed: "Since the evidence for evolution is so absolutely, totally overwhelming, nobody who looks at it could possibly doubt it—if they were sane and not stupid." Referring to his popular book *The God Delusion* (which has sold more than a million copies), Dawkins admitted that it "is my long-expected, long-worked-on full frontal attack on religion. To me, science . . . and religion . . . [are] about trying to explain existence. It's just that religion gets the wrong answer."[20] Dawkins also admitted that "when I discovered Darwinism, I realized there was this magnificently elegant . . . explanation. I didn't quite understand at the beginning; when I did understand, then that finally killed off my remaining religious faith."[21]

There are other outspoken critics of religion. Dr. Will Provine, Professor of History of Biology at Cornell University, once spoke of his first evolution class and how after studying the text, he immediately gave up on deity. He said, "No God, no life after death, . . . no ultimate meaning in life, and no human free will, are all deeply connected to an evolutionary perspective. You're here today and gone tomorrow. And that's all there is to it."[22]

What a discomforting thought!

Eugenie Scott, executive director of the National Center for Science Education which claims to be "the premier institution dedicated to keeping evolution in the science classroom,"[23] wrote: "The United States stands out among developed countries in its low acceptance of one of the major organizing principles of science [evolution]." She continued, "because of its deep religious and historical roots, creationism will not go away any time soon."[24]

With these attacks on religious faith, is it any wonder that believers are reluctant to accept the whole theory of evolution? Evolution calls into question cherished religious beliefs. For instance, if there was no Fall of Adam, as evolution supposes, there would be no need for an Atonement. Also, if man is just a product of chance events, there would be no purpose-driven creation. And if there is no purpose-driven creation, the idea of a divine Creator who is full of justice and mercy can be called into question.

Nephi was very direct in teaching about the dangers of getting caught up in one's own understanding:

> O that cunning plan of the evil one! O the

> vainness, and the frailties, and the foolishness of men! When they are learned they think they are wise, and they hearken not unto the counsel of God, for they set it aside, supposing they know of themselves, wherefore, their wisdom is foolishness and it profiteth them not. . . . But to be learned is good if they hearken unto the counsels of God. . . . And whoso knocketh, to him will he open; and the wise, and the learned, and they that are rich, who are puffed up because of their learning, and their wisdom, and their riches—yea, they are they whom he despiseth; and save they shall cast these things away, and consider themselves fools before God, and come down in the depths of humility, he will not open unto them. (2 Nephi 9:28-29, 42)

The Prophet Moroni gave us the key for identifying deceptive ideas when he wrote:

> But behold, that which is of God inviteth and enticeth to do good continually; wherefore, everything which inviteth and enticeth to do good, and to love God, and to serve him, is inspired of God.
>
> But whatsoever thing persuadeth men to do evil, and believe not in Christ, and deny him, and serve not God, then ye may know with a perfect knowledge it is of the devil. (Moroni 7:13, 17)

Evolution's Far-reaching Influence

It is interesting that the theory of evolution was not inferred from solid scientific evidence in fields like molecular biology, paleontology, and embryology. Rather it was more or less assumed to always be true.[25] This assumption influences the way it is being presented to society. Phillip Johnson observed that because scientists accepted evolution "before it was rigorously tested," many have sought "to convince the public that naturalistic processes are sufficient to produce a human from a bacterium, and a bacterium from a mix of chemicals." Ever since then, Johnson continued, "Evolutionary science became the search for confirming evidence, and the explaining away of [the] negative."[26]

Rather than focusing on the evidence (or lack of evidence) for evolution, Darwinists would like the rest of us to just make up our minds that their theory is true. This sentiment is evident in the words of Susan Bury who wrote: "It's time we finally wring out the public debate over whether evolution occurred and gave rise to the living world. Evolution is real, and we are among its products. Let's get over it [i.e., the aversion to evolution], get with it, and get on with it."[27] Some LDS scholars apparently share her view. David Bailey wrote:

> I must confess that I find the continuing debates about what some LDS general authorities have said about evolution and the age of the earth, or what the "official" LDS position is on these issues to be rather tiresome at this point in time. Such debates just don't matter any more, except perhaps to historians, because *the scientific facts supporting the basic notions of these theories are now just too well established.* Past writings by General Authorities or other religious leaders criticizing these theories only prove that they are human. I think it's time we accepted this and moved on.[28] (emphasis added)

It seems odd to presume that "the scientific facts supporting the basic notions of these theories are now just too well established." We are being led to believe that Darwinism is more secure than the facts warrant. If evolution has achieved consensus status in the world, does that mean that Latter-day Saints ought to join in so that we feel like we belong? What happened to the old Mutual slogan, "Dare to be different"? It is sad to think that some Latter-day Saints would accept a theory that runs contrary to scriptures and divine teachings from General Authorities just because others say it is true.

Some members have concluded that with the accumulation of data over the past few decades, the evidence in favor of evolution is overwhelming and its truthfulness cannot be denied.[29] Thus they see evolution as something that needs to be reconciled with Church doctrine. In trying to reconcile evolution and LDS doctrine, they suppose that evolution is the means by which God accomplished the creation (called theistic evolution). They persist in evolution with the gospel even though some aspects of the theory have been ruled out by official statements from the Church (see Chapters Seven and Eight on teachings of the prophets regarding evolution).

Moving beyond the sphere of Latter-day Saints, Darwinism has had an immense influence on society as a whole. David Berlinski noted that many people now believe "that human behavior is 'the product of evolution'" and that it "is now more than a theory, it is a popular conviction."[30] This thinking has far reaching ramifications, as evidenced by recent attempts by one author to rationalize the actions of a popular athlete who had been unfaithful to his wife. The author seemed to excuse the athlete's infidelity by saying, "Evolution has shaped men to be sexual opportunists."[31] If the author is right, then evolution justifies immoral behavior, and there are no consequences for infidelity besides the misery it causes betrayed family members.

Conclusion

It is a fact that things change over time, and that diversity occurs in new generations; however, the supposed changes from chemicals to living things and from the simple to complex—without God's influence—have not been, and cannot be, demonstrated. Unless there is intelligent intervention, things go in the opposite direction, from order toward disorder, and from complex to simple.

Macbeth aptly noted that:

> The creationists freely admit that they are relying on revelation and that their view requires an act of faith. The evolutionists ought to admit that they are relying largely on speculations and that speculations should not be taught as hard science.[32]

The information in this chapter suggests that there are abundant problems with some aspects of the theory of evolution—at least in the way it is commonly taught. Once the details are revealed, evolution and creation will cease to be in conflict. Until then, we are left with a great deal of uncertainty. In the meantime, believers should not lose their faith in God and His involvement in the creation because of a very popular theory!

Notes

[1]Jacob 4:8–13; 1 Nephi 8:19–30; 2 Nephi 9: 28–29, 42; 2 Nephi 28:1–6; 2 Nephi 32:4; Jacob 7:11–20; Mosiah 12:27; Alma 12:9–11, 28-30; Alma 30:13–18; Alma 32:34; Alma 36:26; Helaman 16:15–18; Deuteronomy 29:29; John 3:16–21; 1 Corinthians 1:18–31; 1 Corinthians 2; Colossians 2:8. Glen L Pearson, and Reid E. Bankhead, *Building Faith with The Book of Mormon*, Provo, Utah. Joseph Educational Foundation 1994, p. 69–71. I have added: Doctrine and Covenants 101:32–34; and Isaiah 55: 8–9.

[2]1 Nephi 9:6; Mosiah 3:19; Mosiah 16:3–8; Alma 12:21–27; Alma 22:13; Alma 42:2–7; Helaman 14:16; Mormon 9:12–13; Ether 3:2; Romans 5:14-20; Moses 4:-6-32; Moses 5:4-11; Moses 6 46–62; Doctrine and Covenants 20:18–29 and 29: 40–42. Pearson & Bankhead, *Building Faith*, 47–48.

[3]Elder Widtsoe was a scientist, as well as a member of the Quorum of the Twelve Apostles from 1921 until his death in 1952.

[4]John A. Widtsoe, *Evidences and Reconciliations: Aids to Faith in a Modern Day* (Salt Lake City: Bookcraft, 1943), 149.

[5]David Salkeld, "Intelligent Design?" *Chronology & Catastrophism Review*. Society for Interdisciplinary Studies (2009), 2.

[6]Caroline Crocker, *Free to Think: Why Scientific Integrity Matters* (Southworth, Washington: Leafcutter Press, Publisher, 2010), 30.

[7]Shankar Vedantam, "Eden and Evolution." http://www.washingtonpost.com/wp-dyn/content/article/2006/02/03/AR2006020300822_pf.html.

[8]Norman Macbeth, "How to Defuse a Feud." *Kronos: A Journal of Interdisciplinary Synthesis*, VII 4, (1982), 3.

[9]National Academy of Sciences. *Science and Creationism: A View from the National Academy of Sciences*, 2nd Edition. (Washington DC: National Academy Press, 1999), 10.

[10]Juliana Boerio-Goates, "The Law of Increasing Disorder." In *Physical Science Foundations*, 2nd Edition, ed. Jennifer Berry (Provo, Utah: BYU Academic Publishing. 2006), 214.

[11]Hugh Nibley, *The Ancient State* (Salt Lake City: Deseret Book, 1991), 450, 453–454.

[12]Hugh Nibley, *Temple and Cosmos: Beyond this Ignorant Present* (Salt Lake City: Deseret, 1992), 4–6.

[13]Hugh Nibley, *Old Testament and Related Studies* (Salt Lake City: Deseret Book, 1986), 24.

[14]Nibley, *Old Testament and Related Studies*, 24.

[15]Jonathan Wells, "Common Ancestry on Trial," In *Darwin's Nemesis*. Edited by William A. Dembski, (Downers Grove, IL, 2006), 6.

[16]Karen Armstrong, "Two Paths to the Same Old Truths," *New Scientist*, July 30, 2005, 42.

[17]"Expelled: No Intelligence Allowed." Premise Media Corp. DVD. 2008, 1:02:06, 1:03:54.

[18]"Expelled" 1:03:21.

[19]C. D. Darlington. "The Origin of Darwinism." *Scientific American* (May, 1959), 60.

[20]"Expelled" 30:50, 54:38.

[21]"Expelled" 1:02:29.

[22]"Expelled" 58:03.

[23]http://ncse.com/.

[24]Eugenie C. Scott, "Not (Just) in Kansas Anymore." *Science*, May 5, 2000: 813–814.

[25]Jonathan Wells. "Common Ancestry on Trial," In *Darwin's Nemesis*, ed. William A. Dembski, (Downers Grove, IL, 2006), 6.

[26]Phillip Johnson, *Darwin on Trial* (Downers Grove, Ill InterVarsity Press, 1991), 150.

[27]Susan Bury, "How to Live with Evolution," *Skeptical Inquirer*, March/April 2001, 56.

[28]D. H. Bailey, "Mormons and Evolution." http://www.tungate.com/LDS_scientist.htm.

[29]Trent D. Stephens and D. Jeffrey Meldrum. *Evolution and Mormonism: A Quest for Understanding* (Salt Lake City, Signature Books, 2001), xviii.

[30]David Berlinski, *The Devil's Delusion: Atheism and Its Scientific Pretensions* (New York, Basic Books, 2009), 166.

[31]Andrew Trees, "Is Tiger Woods' Marriage Over." *The Early Show*, CBS, December 16, 2009.

[32]Macbeth, "How to Defuse a Feud," 1–2.

Chapter 6

Layers of Theistic Evolution

Dave S. Collingridge

Evidence for microevolution is all around us. We see it in the way animals from the same species vary in shape, size, and color. If microevolution is so obvious, why has Darwinism generated so much controversy? The controversy largely stems from Darwin's claim that evolution is capable of changing one life form into a completely different life form (i.e., macroevolution). Along with this, Darwin also suggested that complex species evolved from less complex species, and that living things share a common ancestor.

Historically, the idea that all life, including humans, evolved from lower life forms, has been problematic for people who believe in God. Believers don't like being told their distant ancestors were hominids. This idea violates the Biblical account of God creating Adam and Eve and placing them on the earth. It comes as no surprise then that believers have good reason to reject Darwinism.

At the same time, other believers accept Darwinism. They've reconciled Darwinism with their religious beliefs.

Reconciling a belief in a Supreme Creator with Darwinism is known as theistic evolution (TE). Theistic evolution has grown in popularity among Christians. Its growing influence counts as one of the major successes of Darwinism in recent years. Theistic evolution is becoming popular because it enables people to retain both their religious and evolutionary beliefs. There is no longer a need to view one as true and the other as false; both are accepted as true.

As we analyze TE's claim that evolution and the gospel are compatible, we'll find that there is no reliable path to reconciliation. Religion and evolution hold tenets that directly oppose one another. In order for TE to work, changes must be made to either evolution or religion. A viable reconciliation requires major concessions from one belief system or the other, thus leaving neither fully intact.

Unguided Theistic Evolution

Unguided theistic evolutionists believe in both God and evolution. They believe that when the Lord created the earth, He provided an environment that could sustain life and allow it to evolve. After creating a world on which life could flourish, He stopped being actively involved.

The Lord was no longer involved because He knew evolution would create life for Him. He knew single-celled organisms would evolve into complex life forms and, given enough time, would eventually evolve into humankind. In a manner of speaking, after creating the earth, the Lord merely "watched" as the evolutionary process unfolded.

Unguided TE is popular among people who want to embrace evolution without abandoning their faith. At first glance, it seems to serve this purpose quite well. It allows people to accept that living things evolved through blind and undirected processes, while at the same time preserving a belief in a Supreme Creator. But their acceptance of both God and evolution is not problem-free. Unguided TE offers a less than ideal compromise between religion and evolution. It is not the perfect harmony people have hoped for.

Physical Forms

According to unguided TE, the Lord was not directly involved in the creation of complex life forms. In fact, because He turned over the creation to undirected evolutionary processes, the Creator never really intended to create anything at all, including humans. The creation of frogs, dogs, and humans just happened by chance. This view is evident in the words of one unguided theistic evolutionist who wrote, "mankind's appearance on this planet was not preordained." He continued, "that we are here . . . is an afterthought, a minor detail, a happenstance in a history that might just as well have left us out." He added, if the creation were allowed to play out again, "events might come out differently at every turn."[1]

In other words, if the creation played out again with all its unintended consequences, the world would be a very different place. Animal and plant life would look very different. Humans would look very different. With unguided TE, humanity does not have a predestined shape. Physical form does not matter. Our bodies can take on many different shapes. All that matters is that humankind's physical form is capable of housing spirits. As one unguided theistic evolutionist put it, our spirits could have just as easily been put into the bodies of a sentient dinosaur or a super intelligent mollusk (e.g., a slug).[2] Supposedly God would be fine with putting our spirits into a dinosaur or mollusk, as long as we are capable of self-awareness and of knowing and loving Him.[3]

An obvious problem with being a dinosaur or mollusk is that these life forms are not created in the image of God. We know from gospel teachings that the Father has a body similar to ours. We are created in His image. Being created in His image does not just mean being capable of self-reflection and love. It means we actually look like Him.

With the infinite number of twists and turns that unguided evolution can take, left on its own, it would never produce physical bodies in the image of God. Only a directed and purposeful creation can achieve the proper physical shape. The proper physical shape is a bipedal structure with two arms, one head, two eyes—and no antennae! The Lord didn't intend for us to stomp around like a Tyrannosaurus Rex or slither around like a slug, no matter how much fun doing those things might have been.

Return of the Demiurge

Unguided TE is not a new concept. Author John G. West pointed out that unguided TE represents a resurgence of an ancient, fictional creation story. This fictional creation story was popularized by a belief system known as Gnosticism.[4]

Gnosticism was a first century AD religious movement that got its name from the Greek word gnosis, meaning "knowledge." Gnosticism is used to describe this religion because its followers believed that salvation was achieved by acquiring special knowledge. With knowledge as its guiding principle of salvation, Gnostics tended to downplay the importance of Jesus' Atonement.

Gnostic views on redemption apparently infiltrated some of the early Christian churches. The apostles were aware of these heretical views and took efforts to stop them from spreading. For instance, the Apostle Paul warned Timothy to "guard what has been entrusted to you." In the next sentence he clarified what he meant by guarding what had been entrusted to him when he urged Timothy to "Avoid the godless chatter and contradictions of what is falsely called knowledge [gnosis], for by professing it some have missed the mark as regards the faith" (1 Tim. 6:20–21).

Another heretical gnostic belief was that all matter, including human flesh, is corrupt. Because matter is corrupt, the material world is evil and to be shunned. The Gnostics surmised that because the material world was corrupt, it could not have been created by God. This view created problems for them. If the world was not created by the Lord, then who created it? To solve this problem, the Gnostics borrowed the concept of the Demiurge from ancient Greek philosophy.

In ancient Greek philosophy, the Demiurge was a craftsman who created the physical world. The image of a world-creating craftsman suited the Gnostics. In the Demiurge they found an entity capable of creating a world independent of God's influence. Early Christian leaders were aware of the concept of the Demiurge. Hippolytus of Rome, a third century AD Christian theologian, commented on the unorthodox nature of the Demiurge. He wrote, "For the Demiurge, they say, knows nothing at all, but is, according to them, devoid of understanding, and silly, and is not conscious of what he is doing or working at. . . . [He] himself imagines that he evolves the creation of the world out of himself."[5]

Hippolytus' description of the Demiurge bears a striking resemblance to the creative force in unguided TE. Like the Demiurge, the creative force in unguided TE knows nothing at all, is devoid of understanding, and is not conscious of what it is doing. Also, like Gnosticism which claimed that the Demiurge created life without divine involvement, unguided TE claims that evolution created all living things independent of God's influence.

Contrary to unguided TE, the Lord did not leave the creation of life to the vicissitudes of unguided, natural processes. The creation of all living things was purposeful and directed. The Lord orchestrated the creation of life by the power of His spirit, and by the power of His spirit, He gave life and sustenance to all living things on earth (Mosiah 4:9).

Guided Theistic Evolution

Another brand of evolution that is growing in popularity is guided theistic evolution. Guided TE is the belief that the Lord guided evolutionary processes to bring about the creation of all living things, including humans. In a manner of speaking, He used Darwinian processes to create the complex life we find all around us today.

Guided TE is popular with some Christians because it allows for evolution while preserving a belief in the Lord as an active Creator. People who accept guided TE are in distinguished company. They've joined the ranks of respected scientists such as natural selection's co-founder Alfred Wallace (1823–1913), eminent Harvard botanist Asa Gray (1810–1888), and Human Genome Project director Francis Collins.

At first glance, it looks as though guided theistic evolutionists have the best of both worlds. They have the respect of the scientific community for accepting evolution. At the same time, they've preserved their belief in an active Creator. But a closer look reveals that guided TE has put them in a difficult situation. In order to work, guided TE requires concessions from both evolution and Christianity. Evolution must be changed to make it fit with an active Creator, and Christianity must be changed to make it fit with evolution. Let's take a closer look at why guided TE is a less than ideal compromise.

Omnipotent Creator

The realities of an omnipotent Creator represents a major stumbling block for guided TE. If the Lord used evolution to create humanity, He would have accomplished this feat by manipulating natural selection. This means that He would have facilitated the evolution of humans by creating environmental conditions that actively promoted the survival or destruction of certain species. This type of involvement would have been necessary to steer evolution down a path that led to humanity.

The image of an omnipotent Creator micromanaging environmental events raises an important question. Why would an all-powerful being limit Himself to intervening in a seemingly countless number of natural events to bring about the fullness of His creation, namely humanity? If the Lord wanted to create something as important as humans, He would do it in a direct and purposeful way. There would be no need for millions of years of minor adjustments. The idea that the Lord resorted to tweaking environmental circumstances over millions of years is not consistent with the realities of an all-powerful Creator.

Adam and Eve

Guided theistic evolutionists generally believe that Adam and Eve's bodies were the end result of a long line of evolutionary processes. If we trace the parentage of Adam and Eve back far enough, we would find hominids, primates, and other lower life forms. These lower life forms supposedly evolved into Adam and Eve.

This view does not fit well with the gospel.

To begin with, the scriptures teach that Adam and Eve were the first human parents to inhabit the earth. If Adam and Eve evolved, they would not have been the first parents.

Their parents or some other ancestor would have been the first human parents.

The scriptures also teach that humans did not die before the Fall. This means that Adam and Eve could not have had ancestors who suffered death. If we accept no human death before the Fall, we must reject the idea of Adam and Eve having mortal parents who died before the Fall.

Also, before the Fall, Adam and Eve could not die. Their bodies were immortal. It seems improbable that a natural process like evolution could create immortal bodies. The creation of immortal bodies certainly requires supernatural intervention.

Random Variation

In the currently accepted Darwinian model, natural selection is capable of changing the dominant traits within a species. However, another natural process is required to produce major morphological change. That other process is random genetic variation.

Random genetic variation is a major source of trait variability. It is capable of producing traits that either increase or decrease an organism's chances of survival. Over long periods of time, the combination of natural selection acting on traits produced by random genetic variation is believed to create new life forms.

If evolution is guided, how do guided theistic evolutionists account for the fact that genetic variation is supposed to be random? "Random" suggests a lack of aim or purpose, yet guided theistic evolutionists believe evolution is guided by the Creator. Some guided theistic evolutionists deal with this dilemma by saying that genetic variations are not really random; they only *appear* random.[6] They believe that what appears random to humans might not really be random to the Lord. Genetic events orchestrated by the Creator only appear random to us humans because they have an element of uncertainty and unpredictability. Supposedly, if we had God's perspective, we would find that genetic variations are really divinely directed events.

Not knowing whether genetic variations are truly random does not change their underlying reality. If a genetic variation is truly random, then it did not arise by design. If a genetic variation is truly designed, then it did not arise by random processes. A genetic variation cannot be both random and directed. It is either one or the other.

If we assume that the Lord purposely manipulated genetic material to bring about the evolution of humanity, then those events would have certainly been directed. Yet according to Darwinism, evolutionary events are necessarily "purposeless,"[7] "undirected,"[8] and "without either plan or purpose."[9] Saying that genetic events were directed is contrary to the commonly accepted principles of evolution.

The debate over whether evolutionary events are directed or random comes down to more than just semantic word play. There are real inconsistencies between guided TE's account of the creation and mainstream evolution. The eminent botanist Asa Gray recognized these inconsistencies. Realizing that his belief in divinely directed evolution represented a major deviation from orthodox evolution, Gray was led to admit that his version of guided evolution was "very anti-Darwin."[10]

State-Sponsored Theistic Evolution

Despite inconsistencies between TE and Christianity, TE is gaining converts. One reason for TE's success is that it has gained a powerful ally. It has been promoted by federally funded science organizations.

Not long ago, the National Center for Science Education

(NCSE) realized that it had a problem convincing Christians to accept evolution. To remedy this problem, it launched an effort to change Christians' perceptions of evolution. It sought to convince believers that evolution is not hostile toward religion, and that evolution and religion are compatible. In short, it promoted TE.

In an effort to win over Christians, the NCSE provided churches with educational materials promoting a reconciliation between religion and evolution, and facilitated the distribution of written statements from religious leaders who have accepted TE.[11] The National Academy of Sciences (NAS) also joined the effort with pro-TE statements of their own, such as that "the evidence for evolution can be fully compatible with religious faith."[12]

Why are atheists in the NAS and NCSE promoting a version of Darwinism that they ultimately disagree with, namely theistic evolution? Why are they making claims like "evolution fits well with good theology,"[13] when so many of them do not even believe in God? There is an ulterior motive at work here.

Prominent atheist and author Richard Dawkins offered an explanation as to why atheists promote TE. He suggested that atheists promote TE because they "desperately [want] to be friendly to mainstream, sensible [Christian] people." He continued, "the way [they] do that is to tell [Christians] that there's no incompatibility between science and religion."[14]

Atheists know they have little chance of convincing Christians to accept godless evolution, but they may be able to convince Christians to accept guided TE. If they can get believers to accept guided TE, then it may be possible to convince them to go one step further and accept unguided TE. Once believers have accepted unguided TE, the divide that separates them from godless evolution no longer seems so wide. They merely need to cast off the role of deity in the creation altogether to arrive at unguided, purposeless creation – the same brand of creation embraced by atheists. Thus we see how, with a foot in the door, TE can become a stepping stone to accepting a brand of evolution that completely rejects divine influence.

Conclusion

The Savior taught, "No man can serve two masters" (Matthew 6:24). In the context of the creation, this means we cannot buy into evolution while holding fast to the scriptures. They are too much at odds.

The Prophet Joshua taught that the decision of what to believe is up to us (Joshua 24:15). We can choose to embrace the doctrines of evolution or embrace the doctrines of the gospel. Hopefully we will choose to follow Joshua's example and likewise declare: As for me and my house, we will accept the revealed word of the Lord.

Notes

[1]Kenneth R. Miller, *Finding Darwin's God: A Scientist's Search for the Common Ground between God and Evolution* (New York: Harper-Collins, 2007), 272.

[2]Kenneth R. Miller, *Finding Darwin's God*, 252.

[3]Kenneth R. Miller, *Finding Darwin's God*, 252.

[4]John G. West, "Nothing New Under the Sun," in *God and Evolution*, ed. Jay Richards (Seattle: Discovery Institute Press, 2010), Chapter 1.

[5]Hippolytus, *The Refutation of All Heresies*, Book VI, Chapter XXVVIII.

[6]See, for example, Francis S. Collins, *The Language of God: A Scientist Presents Evidence for Belief* (New York: Free Press, 2006), 205.

[7]Salvador E. Luria, Stephen J. Gould, & Sam Singer, *A View of Life* (San Francisco: Benjamin-Cummings, 1981), 574.

[8]Kenneth R. Miller & Joseph S. Levine, *Biology* (New Jersey: Prentice Hall, 2000), 658.

[9]Kenneth R. Miller & Joseph S. Levine, *Biology*, 658.

[10]Asa Gray, in Daniel C. Gilman, *The Life of James Dwight Dana* (Charleston, SC: BiblioBazzar, 2009), 302.

[11]John G. West, "Having a Real Debate," in *God and Evolution*, ed. Jay Richards (Seattle: Discovery Institute Press, 2010), 53–54.

[12]Ralph J. Cicerone, Harvey Fineberg, & Francisco J. Ayala, *Science, Evolution, and Creationism: A View from the National Academy of Sciences* (Washington, DC: National Academy of Sciences, 2013), Preface.

[13]Michael Shermer, "Darwin on the Right: Why Christians and Conservatives Should Accept Evolution," *Scientific American* (October 2006).

[14]Richard Dawkins, quoted in *Expelled: No Intelligence Allowed*, Premise Media, 2008.

Chapter 7

Teachings of the Prophets: Restoration to First Half of the 20th Century

Ryan Nilsson

We have no right to take the theories of men, however scholarly, however learned, and set them up as a standard, and try to make the Gospel bow down to them; making of them an iron bedstead upon which God's truth, if not long enough, must be stretched out, or if too long, must be chopped off – anything to make it fit into the system of men's thoughts and theories! On the contrary, we should hold up the Gospel as the standard of truth, and measure thereby the theories and opinions of men. —Orson F. Whitney

A brief perusal of the teachings of the earlier prophets of the Restoration—those of the First Presidency and the Quorum of the Twelve – will show that there has been a very consistent voice on the matter of evolution. At no time have the prophets taught that evolution—the theory of man evolving from lower animals—is true. But rather they have been consistent in denouncing that aspect of evolution as false.

Spencer W. Kimball assured us that "there will never be a majority of the Council of the Twelve on the wrong side at any time. The Lord has chosen them. . . . And those people who stand close to them will be safe."[1] A short tour of the words of the prophets on the matter of evolution will here be made, to demonstrate to readers their teachings on this subject.

The Days of the Prophet Joseph Smith

Joseph Smith died on June 27, 1844, and so he was not contemporary with the theory of evolution, since Darwin's *Origin of Species* was published in 1859. Had the prophet been contemporary with *Origin of Species*, there is little doubt that he would have commented on it and contradicted its claims. The Prophet Joseph, after all, had the benefit of the wide visions of eternity, and he knew the truth by visionary experience – and yet he was a strict scripturalist.

Recall that according to Darwinism, all life forms evolved from a single-celled organism, which slowly branched out into different kinds of creatures through the ages. The Prophet Joseph Smith taught us in no uncertain terms that life does not do that. It does not transcend the general "kinds" which God has created in the beginning. He said,

> God has made certain decrees which are fixed and immovable; for instance, God set the sun, the moon, and the stars in the heavens, and gave them their laws, conditions and bounds, which they cannot pass, except by His commandments;

> they all move in perfect harmony in their sphere and order, and are as lights, wonders and signs unto us. The sea also has its bounds which it cannot pass. God has set many signs on the earth, as well as in the heavens; for instance, the oak of the forest, the fruit of the tree, the herb of the field, all bear a sign that seed hath been planted there; *for it is a decree of the Lord that every tree, plant, and herb bearing seed should bring forth of its kind, and cannot come forth after any other law or principle.*[2]

The prophet also taught that life does not come from non-living matter, as many evolutionists suppose. According to one theory of abiogenesis, also called spontaneous generation, the first single-celled organism arose from non-living matter in a "warm little pond." But the prophet taught us, "Where was there ever a son without a father? And where was there ever a father without first being a son? Whenever did a tree or anything spring into existence without a progenitor? And everything comes in this way."[3]

Clearly life must come from already existing life, and living things cannot transcend the general "kind" of life that they belong to. These two verities taught by the Prophet Joseph Smith invalidate the evolutionary claims of spontaneous generation.

The President of the Church Speaks (Brigham Young and John Taylor)

Since the days of the Prophet Joseph Smith, when Darwin's book exploded onto the scene, the brethren of the Church through the nineteenth century had not been shy about denouncing its theories. It is worth considering that if Darwin's theory was the incredible ray of light and truth that evolutionists claim it was, then why did God's authorized servants so uniformly reject it upon its outset?

For instance, President Brigham Young declared his unalterable opposition to the theory of evolution and lamented that it was being taught in Zion without being balanced by gospel teachings. He said,

> We have enough and to spare, at present in these mountains, of schools where young infidels are made because the teachers are so tender-footed that they dare not mention the principles of the gospel to their pupils, but have no hesitancy in introducing into the classroom the theories of Huxley, of Darwin ... This course I am resolutely and uncompromisingly opposed to.[4]

Also, President Brigham Young taught that,

> The whole Scriptures plainly teach us that we are the children of that God who framed the world. Let us look round and see whether we can find a father and son in this congregation. Do we see one an elephant, and the other a hen? No. Does a father that looks like a human being have a son like an ape, going on all fours? No; the son looks like his father. There is an endless variety of distinction in the few features that compose the human face, yet children have in their countenances and general expression of figure and temperament a greater or less likeness of their parents. You do not see brutes spring from human beings. Every species is true to its kind. The children of men are featured alike and walk erect.[5]

The foregoing statement by President Young clearly acknowledges microevolution (observable variations among the kinds of animals), but denies macroevolution (theoretical changes from one basic kind of animal to a different kind). Without splitting hairs about the meaning of the ambiguous term, "species," or applying a contemporary meaning to Brigham Young's use of it, it is telling that President Young declared, "Every species is true to its kind," and that for evidence of this fact he appealed both to the scriptures and to common experience. In other words, while there are a great variety of dogs, we do not see dogs producing offspring that are not dogs.

One of the most notable denunciations of evolution from the nineteenth century came from John Taylor. While he was President of the Church, President Taylor wrote his famous treatise *Mediation and Atonement.* In that work, President Taylor wrote that "These principles [of prescribed laws in the animal and plant kingdoms] do not change, as represented by evolutionists of the Darwinian school, but the primitive organisms of all living beings exist in the same form as when they first received their impress from their Maker." He added, "As the horse, the ox, the sheep, and every living creature, including man, propagates its own species and perpetuates its own kind, so does God perpetuate His."[6] The meaning of President Taylor's message is clear: macroevolutionary changes across life forms do not happen.

The Witness of Early Apostles

Erastus Snow, called to be an apostle in 1849, was astute enough to notice that evolution's tale of a single cell evolving into mankind is beyond the power of observation and is therefore not demonstrable. Instead, it must be taken on belief, as a matter of "faith." Elder Snow taught,

> Mr. Darwin, and a kindred school of modern philosophers, would fain try to impress upon us their theory of evolutions, and would have us believe we are descendants from, and only a little in advance of our ancestor, the monkey; and that other inferior grades of animals are aspiring to become monkeys; they fail to demonstrate their theories, simply because they are not demonstrateable.[7]

Elder Snow also criticized "the school of modern philosophers" who propound the theory of evolution, and noted that the theory was inconsistent with what we see in the natural world.[8]

Orson Pratt, the great scriptorian of the nineteenth century and one of the original Twelve, denied evolution by the authority of scripture as well. He remarked that "[evolutionists] would make man look for his origin down to the very reptile and the worm that crawls upon the earth," and he denounced "the evolution theory" as "the creation of man's folly and foolishness." He affirmed divine creation when he said that "we learn through the revelations of God that instead of man's coming up from the poor worm of the dirt, he descended from the being who controls the universe by his power."[9]

Elder Moses Thatcher, while speaking words of encouragement to the youth of the Church in 1891, took occasion to warn them of the false philosophy of evolution:

> I would have you view all the great architectural designs that man has made by the inspiration of his Maker and prove to you that man did not originate from the lower animals, and that the creator is greater than that which is Created.[10]

President George Q. Cannon, First Counselor in the First Presidency, related that he had counseled college-bound young men to beware of the false doctrine of Darwinism. He said that we can know it is false because the scriptures show it to be false:

> We are a very imperfect people. We have much to learn. But, as I said, God has revealed some things to us, and we know they are true; and we can bring all the science of the world to that standard which God has given unto us. As I have told many of our young men when they have been going to college, how valuable it is to know that these doctrines of Darwin are incorrect, and that man has not come up from that low condition, through thousands or millions of years, to his present stature and reasoning faculties, etc. It is worth something to know that, to begin with, and not spend time over it.[11]

The First Half of the Twentieth Century

Since the publication of Darwin's *Origin of Species*, Church officers had been consistently denouncing the theory of evolution both on theological and scientific grounds. But it was in the early part of the twentieth century that the Church formally coalesced an official stance against evolution. This official stance takes the form of three statements from the First Presidency.

The first and most notable First Presidency declaration on evolution was given during the presidency of Joseph F. Smith. During 1909 there was widespread discussion and excitement about Charles Darwin's theory of evolution. That year marked the 50th anniversary of the publication of the *Origin of Species*. In response, the First Presidency felt it prudent to formulate a statement representing the attitude of the Church towards evolution theory. This statement is titled the "Origin of Man." It is a fairly long dissertation that expresses some doctrinal foundations on divine revelation and the preexistence before unleashing a poignant declaration on the creation. The paragraphs of greatest interest are quoted below:

> In presenting the statement [on evolution] that follows we are not conscious of putting forth anything essentially new; neither is it our desire so to do. Truth is what we wish to present, and truth—eternal truth—is fundamentally old. A restatement of the original attitude of the Church relative to this matter is all that will be attempted here. To tell the truth as God has revealed it, and commend it to the acceptance of those who need to conform their opinions thereto, is the sole purpose of this presentation.
>
> It is held by some that Adam was not the first man upon this earth and that the original human being was a development from lower orders of the animal creation. These, however, are the theories of men. The word of the Lord declared that Adam was "the first man of all men" (Moses 1:34), and we are therefore in duty bound to regard him as the primal parent of our race. It was shown to the brother of Jared that all men were created in the beginning after the image of God; whether we take this to mean the spirit or the body, or both, it commits us to the same conclusion: Man began life as a human being, in the likeness of our Heavenly Father.

> True it is that the body of man enters upon its career as a tiny germ embryo, which becomes an infant, quickened at a certain stage by the spirit whose tabernacle it is, and the child, after being born, develops into a man. There is nothing in this, however, to indicate that the original man, the first of our race, began life as anything less than a man, or less than the human germ or embryo that becomes a man. Man, by searching, cannot find out God. Never, unaided, will he discover the truth about the beginning of human life. The Lord must reveal Himself or remain unrevealed; and the same is true of the facts relating to the origin of Adam's race—God alone can reveal them. Some of these facts, however, are already known [i.e. from the scriptures], and what has been made known it is our duty to receive and retain.
>
> The Church of Jesus Christ of Latter-day Saints, basing its belief on divine revelation, ancient and modern, proclaims man to be the direct and lineal offspring of deity.[12]

This declaration is still considered to be the Church's quintessential statement on the matter of evolution, both by the faithful and the unimpressed. It is hard to deny such a forthright statement from the First Presidency. In fact, as recently as February 2002, the Church declared that this statement still "expresses the Church's doctrinal position" on "the theories of evolution."[13]

Evolutionists in the Church go to great lengths to make it appear as though the First Presidency's 1909 "Origin of Man" statement does not say what it clearly appears to say. They come up with strange and convoluted interpretations of what the "Origin of Man" really says. But we must keep in mind that the First Presidency issued this declaration to the general Church, and therefore chose to put it into words which are readily understandable to the average person. There is no evidence of a special code language being utilized. Rather, the message in the statement is simply and clearly contrary to the theory of evolution.

President Joseph F. Smith's teachings against evolution were not limited to the 1909 "Origin of Man" statement. At the funeral of Rachel Grant (Heber J. Grant's mother), President Smith said, "We did not spring from spawn. . . . We did not pass through the ordeals of embodiment in the lesser animals in order to reach the perfection to which we have attained in manhood and womanhood, in the image and likeness of God."[14] Also, in 1911 he wrote to the young people of the Church that evolution is a "theory which we believe is more or less a fallacy," and noted that "we have abundant evidence that many of those who have adopted in its fullness the theory of evolution have discarded the Bible, or at least refused to accept it as the inspired word of God."[15]

President Joseph F. Smith's declarations against evolution have stood the test of time and have not been undone by scientific discovery. His statements are just as relevant today as they were over a century ago. Speaking on Joseph F. Smith's insight on the doctrines of the gospel, President Gordon B. Hinckley declared:

> It is my opinion that no man, save the Prophet Joseph only, has had a greater and better understanding of the origin and history of the Church, and of its doctrines (not only concerning this life, but also concerning the eternities). There have been other great exponents, but I think none has

> had a broader or deeper understanding, or spoken so eloquently concerning these matters.[16]

It is interesting to note, therefore, that Joseph F. Smith ranks among one of the strongest witnesses against the theory of evolution that the Church produced from the Restoration through the early 20th century.

That Which Is Demonstrated, We Accept With Joy

Not long after the "Origin of Man" statement was released, the First Presidency commented on the Church's approach to scientific theories. In a Christmas message in 1910, the First Presidency made a statement that, while not specifically referencing evolution, had obvious implications for the theory. The First Presidency stated,

> Our religion is not hostile to real science. That which is demonstrated, we accept with joy; but vain philosophy, human theory and mere speculations of men, we do not accept nor do we adopt anything contrary to divine revelation or to good common sense. But everything that tends to right conduct, that harmonizes with sound morality and increases faith in Deity, finds favor with us no matter where it may be found.[17]

The preceding statement proclaims that our religion accepts demonstrated truths of science. Demonstrated truth includes microevolution which refers to variation within the same kinds of animals and is something we see regularly in the world around us. But macroevolution and the idea that all life evolved from a single-celled organism (i.e., common descent), have not, and in all likelihood cannot, be demonstrated. More importantly, because macroevolution is contrary to the teachings of the prophets and scriptures, we cannot accept it in good conscience. This 1910 Christmas message represents the second First Presidency statement against evolution.

The "Mormon" View of Evolution

The third First Presidency statement against evolution appeared during the administration of Heber J. Grant, Joseph F. Smith's immediate successor. In 1925, a landmark legal trial that became known as the "Scopes Monkey Trial" took place in Tennessee. The Scopes trial was intended to resolve the issue of whether evolution should be taught in Tennessee's public schools. The national excitement that the trial generated was sufficient to elicit another statement from the First Presidency (consisting of Heber J. Grant, Anthony W. Ivins, and Charles W. Nibley) on the subject of evolution. Entitled "The 'Mormon' View of Evolution," this statement reads like a shorter version of the previous 1909 "Origin of Man" statement, except that the more explicit paragraphs are not present.

"The 'Mormon' View of Evolution" reinforces that Adam was the first man, a living soul (i.e. a real person), and that mankind is descended from God, not animals. The basic premise of the document, as evidenced by its title, is that Mormons accept the idea of evolving into something greater, but only if understood in the sense of reaching our exaltation (i.e. graduation from a man to a god). The document reads:

> Man is the child of God, formed in the divine image and endowed with divine attributes, and even as the infant son of an earthly father and

> mother is capable in due time of becoming a man, so that undeveloped offspring of celestial parentage is capable, by experience through ages and aeons, of evolving into a God.[18]

How did authorities of the Church at the time interpret "The 'Mormon' View of Evolution"? No doubt spurred by the Scopes Monkey Trial and energized by the authority and spirit of the First Presidency's "'Mormon' View of Evolution," George Albert Smith, then a senior apostle, proclaimed in General Conference,

> I am grateful that in the midst of the confusion of our Father's children there has been given to the members of this great organization a sure knowledge of the origin of man, that we came from the spirit world where our spirits were begotten by our Father in Heaven, that he formed our first parents from the dust of the earth, and that their spirits were placed in their bodies, and that man came, not as some have believed, not as some have preferred to believe, from some of the lower walks of life, but our ancestors were those beings who lived in the courts of heaven. We came not from some menial order of life, but our ancestor is God our Heavenly Father.[19]

Later, as President of the Church, George Albert Smith repeated much the same sentiment, confirming that we are not the product of evolution. He said, "We are the children of God. He is the Father of our spirits. *We have not come from some lower form of life*, but God is the Father of our spirits, and we belong to the royal family, because He is our father."[20]

After referring to the Scopes Monkey Trial in Tennessee, Elder Orson F. Whitney conveyed that "I believe that when

God made man in His own image, He made a man, and not a monkey, nor any other animal out of which man has evolved."[21] In other words, man was created in God's image, rather than coming from lower forms of life as suggested by Darwin's account.

Other Apostolic Statements

Elder John W. Taylor, son of President John Taylor, likewise spoke out against evolution in the October General Conference of 1903. Referring to false doctrines that were creeping into the Church among a limited number of members, he said, "we should not waste our time on . . . evolution, and such things. We have believers in evolution in our midst, and there is a little 'blight upon the top of the trees' in this regard." He then concluded that "All the evolution in the world will not save a single soul; neither will all the Christian Science, neither will spiritualism, nor hypnotism, nor any other ism that is not of God."[22]

During the 1918 General Conference of the Church, Elder Rudger Clawson was teaching on the first chapter of the Bible. He noted the following implications of Genesis chapter 1:

> Now, it seems to me that there is to be a very great lesson learned from this important chapter in the Bible [Genesis chapter 1]. In the first place, I think that it answers perfectly the false doctrine of evolution. We are distinctly given to understand that every living creature was made after its kind. We are not to expect that a lion will grow into a horse, or that a cow will grow into an elephant, but we have reason to believe that a horse will always be a horse. You may be able to improve the horse, but it will still be

> a horse, and so with the other animals of God's creation.[23]

Note Elder Clawson's acknowledgment of microevolution (i.e. "You may be able to improve the horse,") but his denial of evolution across life forms (i.e., "A horse will always be a horse."). This is the reasonable view that many believers hold. We can observe variation within kinds of animals (microevolution), but the idea that animals can evolve into completely different kinds is not scriptural and is not observed in nature.

Elder George F. Richards agreed. While giving a talk in General Conference he said, "There is just the one individual from beginning to end. The fallacy of the reincarnation of man theory is here made apparent, as also the doctrine of evolution, in which it is taught that man originated with the lower animals, the very lowest type, and came up by degrees."[24]

Misrepresented Views of James E. Talmage and John A. Widtsoe

It is common for evolutionists in the Church to claim that notable apostle-scientists James E. Talmage and John A. Widtsoe are on their side, and that these men were certified evolutionists. The evidence does not support such claims. While both believed in an old earth and death before the Fall of Adam, neither of them came close to accepting evolution. In fact, both were harsh critics of the theory in several respects.

It is evident that Elder Talmage openly acknowledged microevolution,[25] as anyone would, but denied macroevolution.[26] He noted,

> Attempts have been made to show that the

> present state of man is the culmination of a long course of development, higher forms having been evolved from inferior progenitors. Undoubtedly much that is classed under the current title, Evolution, is true. That part which comprises facts is true; and with equal assurance be it said that much of the *theory* of evolutionists is error.[27]

The "facts" which are "true" are demonstrable variations of the microevolutionary sort. The "theory" which is in "error" is the story of man evolving from hominids.

In his 1931 "The Earth and Man" address, Elder Talmage made sure to stress that, "We believe that Adam was a real personage, who stands at the head of his race chronologically. To my mind Adam is a historic personage, not a prehistoric being, unidentified and uncertain." He also declared, "I do not regard Adam as related to—certainly not as descended from—the Neanderthal, the Cro-Magnon, the Peking or the Piltdown man. Adam came as divinely directed, created and empowered, and stands as the patriarchal head of his posterity."[28]

Elder Talmage also made a distinction between truths and falsehoods contained in the theory of evolution. He said, "At best, the conception of the development of man's body from the lower forms through evolutionary processes has been but a theory, an unproved hypothesis. . . . Theories have their purpose and are indispensable, but they must never be mistaken for demonstrated facts. The Holy Scriptures should not be discredited by the theories of men; they cannot be discredited by fact and truth." Finally, Elder Talmage declared, "Man is the child of God. . . . He is born in the lineage of Deity, not in the posterity of the brute creation."[29]

Elder Widtsoe commented upon the theory of evolution in his article series titled *Evidences and Reconciliations*. He wrote,

> Had the proponents as well as the opponents of evolution, as a whole or in part, kept in mind that they were discussing a theory, subject to frequent and fundamental change, the civilized world would have been spared much unseemly behavior. Again the warning: Distinguish clearly between facts and the inferences from facts. Certainly, it is a mistake to accept theories in building faith in anything, from religion to our everyday life pursuits.[30]

An astute student of science, John A. Widtsoe was careful to stress the difference between "facts" and "the interpretations of observed facts." The interpretation of facts should not be confused with the facts themselves.[31] This wise counsel has direct and obvious implications for evolution. For instance, it is a fact that fossils exist. How old they are, how they got where they were found, what their ancestors were, and what their progeny were, are inferences and interpretations, and not facts in and of themselves. The existence of the fossils is a known fact. Nearly all of the other information about them is mere interpretation and guesswork, however well-founded it may appear to be. "No scientific worker worthy of his task attempts to give a scientific explanation a higher standing than that of an intelligent guess, supported by existing facts," Elder Widtsoe maintained.[32]

Elder Widtsoe also discussed the difference between observable variations within the same kinds of animals, and creatures transcending their basic kinds to become completely different life forms. In his article, "To What Extent Should the Doctrine of Evolution Be Accepted?" Elder Widtsoe seems to clearly reject macroevolution. He points out what most every evolutionist is unwilling to admit—that while microevolution is observable and unanimously accepted, macroevolution must be accepted as a matter of belief. Though Elder Widtsoe did not use the terms "microevolution" or "macroevolution" in his article, his writings suggest that he clearly supported such a distinction.

Referring to the theory of common descent, Elder Widtsoe wrote, "Yet, at the best the doctrine of the common origin of all life is only an inference of science. After these many years of searching, its truth has not been demonstrated."[33] Again he reiterated, "The doctrine of the common origin of life on earth is but a scientific theory, and should be viewed as such," and "Such hypotheses or theories, which are often helpful, become dangerous when confused with the facts themselves. There are now many theories of evolution, all subject to the normal scrutiny to which all theories should be subjected; and until their probability is demonstrated, it is well to remain wary of them."[34] He further made this indictment of those who claim that evolution has been proven beyond a reasonable doubt:

> The man, learned or unlearned, who declares the doctrine of the common origin of life on earth to be demonstrated beyond doubt, has yet to master the philosophy of science. The failure to differentiate between facts and inferences is the most grievous and the most common sin of scientists.[35]

By their own admission it is clear that neither Elder James E. Talmage nor Elder John A. Widtsoe was a proponent of Darwin's complete theory of evolution. Both men accepted demonstrated microevolutionary changes, but rejected undemonstrated macroevolutionary changes.

Misrepresented Views of Various Apostles

Let's now take a look at statements by other apostles that have been interpreted as supporting evolution. It is strange that someone would reach this conclusion given that none of these apostles have openly declared evolution to be true. In fact, they've made several statements indicating that evolution is false.

One statement that is sometimes cited as supporting evolution appeared in a 1910 editorial in the *Improvement Era* (the forerunner to the *Ensign* magazine), of which Joseph F. Smith was an editor. The statement briefly recounts some creation scriptures and then reads:

> These are the authentic statements of the scriptures, ancient and modern, and it is best to rest with these, until the Lord shall see fit to give more light on the subject. Whether the mortal bodies of man evolved in natural processes to present perfection, through the direction and power of God; whether the first parents of our generations, Adam and Eve, were transplanted from another sphere, with immortal tabernacles, which became corrupted through sin and the partaking of natural foods, in the process of time; whether they were born here in mortality, as other mortals have been, are questions not fully answered in the revealed word of God.[36]

There are some who consider this statement to be an official admission of the possibility of the evolution of mankind. However, it is not conclusive that the passage was authored by President Joseph F. Smith (even though he was apparently an editor of the column), and it does not reflect what President Joseph F. Smith is known to have taught on the subject of evolution and the origin of Adam.

For President Smith's views on the creation of man, readers may consult his other works cited above. It is also important to note that in 1913, President Smith taught that "Man was born of woman; Christ, the Savior, was born of woman and God, the Father, was born of woman. Adam, our earthly parent, was also born of woman into this world, the same as Jesus and you and I."[37] This statement stands in direct contrast to the uncertainty of the *Improvement Era* editorial.

If Adam was born of a woman then who were his parents? The scriptures confirm that Adam was "born into the world" (Moses 6:59) naturally, but also proclaims that Adam was "the son of God" (Moses 6:22). And President Smith himself, in the official "Origin of Man" statement declared, "The Church of Jesus Christ of Latter-day Saints, basing its belief on divine revelation, ancient and modern, proclaims man to be the direct and lineal offspring of Deity."[38]

There is another situation in which one with apostolic authority left open the possibility of evolution, but did not proclaim evolution as truth. In June of 1933, Stephen L. Richards, then a member of the Council of the Twelve, included in the *Improvement Era* "An Open Letter to College Students" in which he wrote,

> If the evolutionary hypothesis of the creation of life and matter in the universe is ultimately found to be correct, and I shall neither be disappointed nor displeased if it shall turn out so to be, in my humble opinion the Biblical account is sufficiently comprehensive to include the whole of the process.[39]

Of course, Elder Richards was certainly in the minority with his opinion that "the Biblical account is sufficiently comprehensive to include the whole of the process [of evolution]." At no point in his letter did he demonstrate scriptural evidence of this comprehensiveness. Neither is this statement tantamount to a declaration of belief in evolution, rather it is a hypothetical, "what if" scenario of what might happen if evolution were proven true. It is telling that Elder Richards closed his letter with these words: "Never abandon a time-tested thing of worth until you are very, very sure that the new is better. Be not ashamed of faith in God."

Notwithstanding this peculiar allowance by Stephen L. Richards, in 1955, while he was a member of the First Presidency, he made his opposition to evolution abundantly clear in a book called *Where is Wisdom?*[40] In this book he criticized the theory of evolution for almost two dozen pages. (*Where is Wisdom?* drew heavily from another book titled *The Origin of Mankind* by Sir Ambrose Fleming, which was recommended to him by scientist and fellow Apostle Joseph F. Merrill.) President Richards' book was glowingly endorsed by David O. McKay, who wrote the foreword to the book.

Another person that LDS evolutionists try to claim for their camp is the Prophet David O. McKay. President McKay was President of the Church for 19 years (1951–1970). He holds the record for longest-serving person in the First Presidency or the Quorum of the Twelve, and is fondly remembered for his many wonderful and insightful teachings. Is it true that David O. McKay espoused evolution, as LDS evolutionists often claim? What is the basis for such thoughts? In a BYU speech, President McKay said,

> Millikan is right when he says, "Science without religion obviously may become a curse rather than a blessing to mankind." But, science domi-

> nated by the spirit of religion is the key progress and the hope of the future. For example, evolution's *beautiful* theory of the creation of the world offers many perplexing problems to the inquiring mind. Inevitably, a teacher who denies divine agency in creation, who insists there is no intelligent purpose in it, will infest the student with the thought that all may be chance. I say, that no youth should be so led without a counter-balancing thought. Even the skeptic teacher should be fair enough to see that even Charles Darwin, when he faced this great question of annihilation, that the creation is dominated only by chance wrote; "It is an intolerable thought that man and all other sentient beings are doomed to complete annihilation after such long, continued slow progress."[41]

Much has been made of President McKay's use of the descriptor, "beautiful" for the theory of evolution. Some treat it as an endorsement of Darwinism. This seems a stretch. A more likely interpretation is that although evolution may have aesthetic appeal to some, by itself it is spiritually dangerous and fails to answer the great questions of life. Interestingly, President McKay said the same thing about evolution during the April 1968 General Conference (read by his son, David Lawrence McKay), except this time he left out the confusing adjective "beautiful." With "beautiful" removed, the statement reads quite unfavorably towards evolution:

> There is a purposeful design permeating all nature, the crowning event of which is man. Here, on this thought, science again leads the student up to a certain point, and sometimes leaves

> him with his soul unanchored. For example, evolution's ["beautiful" removed] theory of the creation of the world offers many perplexing problems to the inquiring mind. Inevitably, a teacher who denies divine agency in creation, who insists that there is no intelligent purpose in it, undoubtedly impresses the student with the thought that all may be chance. I say that no youth should be left without a counterbalancing thought.[42]

Another circumstance concerning President McKay that may be used to support evolution is a written correspondence between the prophet and Professor William Stokes from the University of Utah in 1957. President McKay wrote to William Stokes that "On the subject of organic evolution the Church has officially taken no position." The prophet added that "The book 'Man, His Origin and Destiny' was not published by the Church, and is not approved by the Church. [Rather] The book contains expressions of the author's views for which he alone is responsible."[43]

Indeed the Church has never issued a statement officially denouncing evolution, yet it has issued statements on the Origin of Man expressing doctrines that are clearly contrary to evolution. And although *Man, His Origin and Destiny* was not an official Church publication, acknowledgement of this fact by David O. McKay is hardly an endorsement of Darwinism. In fact, in another letter written only a few years later on the same subject, President McKay said essentially the same thing but then included these words:

> Evolution is a theory. You say that biologists would agree on the general lines of what happened, although there may be less agreement about just how it happened. While scientific

> people themselves differ in their interpretations and views of the theory, any conflicts which may seem to exist between the theory and the truths of revealed religion can well be dealt with by suspending judgment as long as may be necessary to arrive at facts and at a complete understanding of the truth.[44]

Finally, there was a Professor Sterling McMurrin[45] who claimed that David O. McKay told him in private confidence, "I believe in evolution."[46] Whether this anecdote is accurate or not, of course, is impossible to say. But if President McKay did indeed say that, perhaps he was thinking of the observable processes of microevolution (i.e. the variety produced within the basic kinds of animals) when he espoused that belief.

The idea that President McKay believed in evolution contradicts his publicly stated views. He said, "Charles Darwin . . . as he contemplated the greatest descent or origin in his mind of the human family, his theory . . . came to the point that there will come a time when the human family will end. If that theory is right, all sentient beings known as man cannot live."[47] He also indicated that his opinions on the origin of mankind differed from the prevailing view in academia when he wrote, "It is gratifying to note how leading thinkers today, sincere philosophers and scientists are pleading for more spirituality. *They may reason differently from us about the origin of man*, but they all agree that now is the time that the spiritual part of man must be developed."[48]

President McKay was even more explicit relative to the theory of macroevolution—the full theory of Darwinism in which life evolved from lower life forms. He declared:

> The stern fact of life is that animals, as other liv-

ing things, can grow and produce their kind only in accordance with fixed laws of nature and the divine command, 'Let the earth bring forth the living creatures after his kind, cattle and creeping things, and the beast of the earth after his kind.' (Genesis 1:24)[49]

Notes

[1]Spencer W. Kimball, *General Conference*, April 1951.

[2]Joseph Smith, *Teachings of the Prophet Joseph Smith, ed. Joseph Fielding Smith* (Salt Lake City: Deseret Book Company, 1976), 197–198, emphasis added.

[3]Joseph Smith, *Teachings*, 373.

[4]Dean C. Jessee, *Letters of Brigham Young to His Sons*, (Salt Lake City: Deseret Book Company, 1974), 200.

[5]Brigham Young, *Journal of Discourses*, ed. George D. Watt, et al. 26 vols. (Liverpool: F. D. Richards, et al., 1854–1886), 8:29–30, Delivered 1860.

[6]John Taylor, *Mediation and Atonement* (Salt Lake City: Deseret News, 1882), 165–166.

[7]Erastus Snow, *Journal of Discourses*, ed. George D. Watt, et al. 26 vols. (Liverpool: F. D. Richards, et al., 1854–1886), 19:321–325. Delivered January 20, 1878.

[8]Erastus Snow, *Journal of Discourses*, ed. George D. Watt, et al. 26 vols. (Liverpool: F. D. Richards, et al., 1854–1886), 19:265–279. Delivered March 3, 1878.

[9]Orson Pratt, *Journal of Discourses*, ed. George D. Watt, et al. 26 vols. (Liverpool: F. D. Richards, et al., 1854–1886), 20:62–75. Delivered August 25, 1878.

[10]Moses Thatcher, *Y.M.M.I.A. Conference*, held May 31, 1891, as quoted in Brian H. Stuy, ed., Collected Discourses, 5 vols. (Burbank, Calif., and Woodland Hills, Ut.: B.H.S. Publishing, 1987–1992), vol. 2.

[11]George Q. Cannon, *The Chicago Parliament of Religions*, delivered at the Tabernacle, Sep. 24, 1893 as quoted in Brian H. Stuy, ed., Collected Discourses, 5 vols. (Burbank, Calif., and Woodland Hills, Ut.: B.H.S. Publishing, 1987–1992), vol. 3.

[12]Joseph F. Smith, John R. Winder, and Anthon H. Lund, "The Origin of Man," *Improvement Era*, 13 (November 1909), 75–81.

[13]Preamble to "The Origin of Man," reprinted in *Ensign*, February 2002.

[14]Joseph F. Smith, *Gospel Doctrine* (Salt Lake City: Deseret Book Co., 1977), 25–26.

[15]Joseph F. Smith, "Philosophy and the Church Schools," *Juvenile Instructor*, April 1911, 209.

[16]Gordon B. Hinckley, "Joseph F. Smith Building Dedication," *BYU Speeches*, September 20, 2005.

[17]Joseph F. Smith, John R. Winder, Anthon H. Lund, "Words in Season from the First Presidency," *Deseret Evening News*, December 17, 1910.

[18]Heber J. Grant, Anthony W. Ivins, and Charles W. Nibley, "The 'Mormon' View of Evolution," *The Improvement Era*, September 1925, 1090–1091.

[19]George Albert Smith, *General Conference*, October 1925.

[20]George Albert Smith, *General Conference*, April 1946, emphasis added.

[21]Orson F. Whitney, *General Conference*, October 1925.

[22]John W. Taylor, *General Conference*, October 1903.

[23]Rudger Clawson, *General Conference*, April 1918.

[24]George F. Richards, *General Conference*, April 1933.

[25]As discussed elsewhere in this book, *micro*evolution refers to variation within the different kinds of animals that we can observe and demonstrate, such as the variety of dogs.

[26]Recall that macroevolution refers to the unobserved (and unobservable) change from one kind of animal into another, such as from ape to man.

[27]James E. Talmage, "Fallen, But He Shall Rise Again," *Improvement Era*, 22 (October 1919), 12.

[28]James E. Talmage, "The Earth and Man," *Deseret News*, November 21, 1931, 11–21.

[29]Talmage, "The Earth and Man," 11–21.

[30]John A. Widtsoe, *Evidences and Reconciliations* (Salt Lake City: Improvement Era): 28–30.

[31]John A. Widtsoe, *Evidences*, 139–142.

[32]John A. Widtsoe, *Evidences*, 143–145.

[33]John A. Widtsoe, *Evidences*, 143–145.

[34]John A. Widtsoe, *Evidences*, 143–145.

[35]John A. Widtsoe, *Evidences*, 143–145.

[36]Priesthood Quorum's Table, *Improvement Era*, 13 (April 1910), 570.

[37]Joseph F. Smith, *Deseret News*, December 27, 1913, Sec. III, 7.

[38]Joseph F. Smith, John R. Winder, and Anthon H. Lund, "The Origin of Man," *Improvement Era*, 13 (November 1909), 75–81.

[39]Stephen L. Richards, "An Open Letter to College Students" *Improvement Era*, 36 (June 1933):451–453, 484–485.

[40]Stephen L. Richards, *Where is Wisdom?* (Salt Lake City: Deseret Book Company, 1955), 374-396.

[41]David O. McKay, "Message for LDS College Youth," *BYU Speeches*, October 8, 1952, emphasis added.

[42]David O. McKay, *General Conference*, April 1968.

[43]David O. McKay, Letter to Professor William Lee Stokes on February 11, 1957.

[44]David O. McKay, Letter dated February 3, 1959.

[45]Sterling McMurrin did not accept the authenticity of the Book of Mormon or the Prophet Joseph Smith's testimony of it (see Blake Ostler's "An Interview with Sterling McMurrin," *Dialogue: A Journal of Mormon Thought*, 17, 1984, 24-25). For many readers this will cast a significant cloud of uncertainty over his motives and claims regarding David O. McKay's views on evolution.

[46]Gregory A. Prince, *David O. McKay and the Rise of Modern Mormonism* (University of Utah Press: 2005), 46.

[47]David O. McKay, *Gospel Ideals* (Salt Lake City: Deseret Book Co., 1953), 50.

[48]David O. McKay, *Treasures of Life* (Salt Lake City: Deseret Book Co., 1962), 130, emphasis added.

[49]David O. McKay, "Some Fundamental Objectives of a Church University," *BYU Faculty Workshop*, September 17, 1954. Published in Deseret News, September, 25, 1954.

Chapter 8

Teachings of the Prophets: Late 20th Century to Early 21st Century

Ryan Nilsson

[T]here will never be a majority of the Council of the Twelve on the wrong side at any time. The Lord has chosen them; he has given them specific responsibilities. And those people who stand close to them will be safe. – Spencer W. Kimball

The Testimony of Apostles in the Latter Half of the Twentieth Century

George Q. Morris had been a long-time school friend of David O. McKay's, and eventually, during the Presidency of David O. McKay, he was called into the apostleship. George Q. Morris had the distinction of being the oldest man to have been called into the apostleship, being 80 years old at the time of his call. A man of extensive experience, Elder

Morris spoke out against evolution in General Conference in 1956. After relating an experience he had with a young acquaintance, he remarked:

> I am concerned for all of our young people as they go into the field of higher education and meet all the ideas that are so prevalent, which are in sharp conflict with the revelations of God that we know to be true. I suppose he had been taught something about the origin of man according to the theory of organic evolution. I presume he might have been told what I remember reading in some man's writings, that we would have to look for our origin in some minute life in the ocean, perhaps, or in some amoeba-like organism—the simplest form of life. That, he said, was man's beginning. But we know better than that.[1]

Elder Delbert L. Stapley of the Council of the Twelve is another apostle who denounced the idea that mankind evolved from lower organisms. He made it clear that mankind came from a higher source when he taught:

> The scriptures teach us that we are descended from the lineage of the Gods; therefore, we are created in their likeness and image, in both spirit and body, and we are endowed with like character traits, qualities, and powers. We are not the product of evolution from some lower organism of life as man speculates and would have us believe.[2]

Marion G. Romney was another consistent voice against evolution. Four times in a span of 20 years he spoke out against it in the General Conference of the Church. In 1953 he remarked, "It would please me immensely if . . . we could get away from using the language of those who do not believe in the mission of Adam. I have reference to words and phrases such as 'primitive man,' 'prehistoric man,' 'before men learned to write,' and the like." He then reaffirmed, "There were no pre-Adamic men in the line of Adam. The Lord said that Adam was the first man. (Moses 1:34, 3:7; Doctrine and Covenants 84:16.) It is hard for me to get the idea of a man ahead of Adam, before the first man."[3]

Later in the April 1959 General Conference, after quoting an evolutionist's postulations, Elder Romney commented, "I repeat that in our peculiar beliefs and faith in Jesus Christ, comprising the foregoing and other revealed truths concerning him, we greatly differ from other people of the earth."[4]

Speaking in derision during the April 1969 Conference, Elder Romney said, "Man—so our children are told—is an animal, the product of biological evolution. . . . [T]hey are playthings to be exploited and prostituted for the gratification of sensual and lustful desires." He then suggested the remedy for this false teaching: "If our children are to be sufficiently strengthened to stand against this satanic onslaught, they must be taught and trained in the home, as the Lord has directed."[5]

Finally, as the Second Counselor in the First Presidency, President Romney gave a stirring address in the April 1973 General Conference entitled "Man—A Child of God." In it he taught, "it is the universal law of nature in the plant, animal, and human worlds for reproducing offspring to reach in final maturity the likeness of their parents." He also stated that, "The theory that man is other than the offspring of God has been, and, so long as it is accepted and acted upon, will continue to be, a major factor in blocking man's spiritual growth and in corrupting his morals. . . . The concept that man is a beast relieves him of a sense of accountability and encourages him to adopt the fatalistic attitude of 'eat, drink, and be merry, for tomorrow we die.'"[6]

Scriptural Stalwarts Speak Out Against Evolution

An opponent of evolution that has had a large influence on the Church was Joseph Fielding Smith. Joseph Fielding Smith was the son of Joseph F. Smith and the grandson of Hyrum Smith. He was called as an apostle at the young age of 33 in 1910, served in the Quorum of the Twelve for nearly 60 years, and spent 19 years as its President. He was ordained President of the Church in 1970 at the grand age of 93. He is the longest serving apostle to ascend to the position of prophet.

During his lifetime he was widely recognized as an authority on scripture and doctrine. In fact, in 1938 President Heber J. Grant wrote privately that "I consider [Joseph Fielding Smith] the best posted man on the scriptures of the General Authorities of the church that we have."[7] As a scriptural scholar of such caliber, Joseph Fielding Smith was vigorously opposed to evolution. He repeatedly spoke out against the theory during his long tenure as apostle. His attitude toward evolution can be summed up in the following statements:

> This idea that everything commenced from a small beginning, from the scum upon the surface of the sea, and has gradually developed until all forms of life, the beasts of the field, the fowls of the air, the fishes of the sea, and the plants upon the face of the earth, have all sprung from that

> one source, is a falsehood absolutely. *There is no truth in it*, for God has given us his word by which we may know, and all who are led by the Spirit of God can understand through that Holy Spirit, the truth of these things.[8]
>
> I say most emphatically, *you cannot believe in this theory of the origin of man, and at the same time accept the Plan of Salvation as set forth by the Lord our God. You must choose the one and reject the other, for they are in direct conflict and there is a gulf separating them which is so great that it cannot be bridged*, no matter how much one may try to do so.[9]

In 1954, President Joseph Fielding Smith (then President of the Quorum of the Twelve Apostles) published his landmark work on the subject, *Man: His Origin and Destiny*. This book is a masterful work that systematically rebuts all of the false arguments in evolution using both scripture and science. To this day it remains an effective reference and a powerful standard of truth in opposition to Darwin's ideas. In the words of another prophet, Ezra Taft Benson, "It is . . . apparent to all who have the Spirit of God in them that Joseph Fielding Smith's writings [in *Man: His Origin and Destiny*] will stand the test of time."[10]

Another adamant opponent of evolution was Bruce R. McConkie. Elder McConkie was known for his opposition to the theory, some of which he shared in his popular book, Mormon Doctrine. In that book, after criticizing the theory from multiple angles, he concluded unequivocally, "*There is no harmony between the truths of revealed religion and the theories of organic evolution.*"[11] He also taught that members of the Church who accept such theories are "those whose faith falls short of the divine standard."[12]

There were other denunciations of evolution by Elder McConkie, both in public speaking venues and in written works. For instance, in a notable 1980 BYU speech he included evolution as the second of seven "deadly heresies" that plague some members of the Church.[13] In another speech he labeled evolution as one of the "doctrines that weaken faith and may damn."[14] In a later written work, he referred to evolution as a "false philosophy . . . that leads men away from God and salvation."[15]

In an official article in the Ensign magazine which he was asked to author on the subject of creation, Elder McConkie related the circumstances of the Garden of Eden, the Fall of Man, and the change to a natural state of being. He wrote, "There is no evolving from one species to another in any of this."[16] In his final written work, still unpublished when he died, he referred to evolution as "the false view of many self-designated scientists," noting that, "The tendency among them is to present Darwinian theories as established realities," when in fact they are certainly not.

Finally, Mark E. Petersen, an insightful and prolific writer, was another apostle that spoke out strongly against evolution, and not without knowledge of the details of the theory. Elder Petersen wrote at length in his book *Adam: Who Is He?* about the failure of evolutionists to provide evidence for macroevolution (changes from one basic kind to another) and concluded, "The endless evolutionary voyages into the seas of speculation certainly cannot be allowed to destroy our faith in divine revelation, which revelation completely refutes these hypotheses."[17] A similar lengthy discussion can be found in his book, *Moses: Man of Miracles.*[18]

Presidents and Apostles of the Late 20th Century

Presidents of the Church from the latter part of the twentieth century also spoke out against evolution. For example, in the December 1972 First Presidency Message, President Harold B. Lee related a conversation he had with a seemingly faithful member of the Church who asked him a question about the creation. President Lee related: "She wondered about the creation because she had read the theories of the scientists, and the question that she was really asking was: How do you reconcile science with religion? The answer must be, *If science is not true, you cannot reconcile truth with error.*" President Lee suggested the following remedy for such misguided beliefs: "I say that we need to teach our people to find their answers in the scriptures. . . . And if we hear someone teaching something that is contrary to what is in the scriptures, each of us may know whether the things spoken are false—it is as simple as that."[19]

Another prophet, President Ezra Taft Benson, spoke out against evolution during his Presidency. In 1988 he reprinted several of his earlier apostolic talks in book form. These reprints condemned evolution.

> Now, we have not been using the Book of Mormon as we should. Our homes are not as strong unless we are using it to bring our children to Christ. Our families may be corrupted by worldly trends and teachings unless we know how to use the book to expose and combat the falsehoods in . . . organic evolution.[20]

Earlier, as President of the Quorum of the Twelve, President Benson gave his fervent warning in General Conference, which appeared in book form:

> As a watchman on the tower, I feel to warn you that one of the chief means of misleading our youth and destroying the family unit is our educational institutions. There is more than one reason why the Church is advising our youth to attend colleges close to their homes where institutes of religion are available. It gives the parents the opportunity to stay close to their children, and if they become alerted and informed, these parents can help expose some of the deceptions of men like . . . Charles Darwin.[21]

President Benson's successor, Howard W. Hunter, also warned the Church about evolutionary falsehoods. Speaking against the trend of "modernists" to liberalize the scriptures and cast doubt upon their accuracy or historicity, Howard W. Hunter denounced them and asked rhetorically:

> The Old Testament unfolds the story of the Creation of the earth and mankind by God. Should we now disregard this account and modernize the Creation according to the theories of the modernists? Can we say there was no Garden of Eden or an Adam and Eve? Because modernists now declare the story of the Flood is unreasonable and impossible, should we disbelieve the account of Noah and the Flood as related in the Old Testament?[22]

President Gordon B. Hinckley also disavowed belief in evolution. While he never showed much interest in discussing the matter, He did say,

> I believe in evolution, *not organic evolution*, as it is called, but in the evolution of the mind, the

> heart, and the soul of man. I believe in improvement. I believe in growth.[23]

Notable apostles of the twentieth century who have spoken out against Darwinism include Russell M. Nelson and Boyd K. Packer. As a renowned heart surgeon, who is no stranger to science, biology, and anatomy, Elder Russell M. Nelson has consistently voiced opposition to evolution. He wrote this denunciation of the theory in his 1988 book, *The Power Within Us*:

> To me, such theories [of evolution] are unbelievable. Could an explosion in a printing shop produce a dictionary? It's unthinkable! One might argue that it is within a remote realm of possibility, but even if that could happen, such a dictionary could certainly not heal its own torn pages, renew its own worn corners, or reproduce its own subsequent editions. . . . It is incumbent upon each informed and spiritually attuned person to help overcome such foolishness of those who would deny divine creation or think that mankind simply evolved. By the spirit, we perceive the truer and more believable wisdom of God.[24]

Elder Boyd K. Packer declared in the 1983 October General Conference that "When we understand the doctrine of premortal life, then things fit together and make sense. We then know that little boys and little girls are not monkeys, nor are their parents, nor were theirs, to the very beginning generation."[25]

Elder Packer proclaimed in General Conference one year later that "A bird will not become an animal nor a fish. A mammal will not beget reptiles, nor 'do men gather . . . figs of thistles.'" He went on to say, "In the countless billions of opportunities in the reproduction of living things, one kind does not beget another. . . . The pattern for all life is the pattern of the parentage." He concluded, "Surely no one with reverence for God could believe that His children evolved from slime or from reptiles."[26]

Statements from Brethren in the Twenty-First Century

The twenty-first century has likewise seen its share of statements against evolution. President Boyd K. Packer continued his opposition to evolution theory, repeating some of his earlier objections in a 2006 Women's Conference talk entitled "Children of God." He said:

> No lesson is more manifest than that all living things do as the Lord commanded them in the creation. They reproduce after their own kind (see Moses 2:12, 24–25). They follow the pattern of their parentage. Everyone knows that. Every four-year-old knows that! (Matthew 7:16).
>
> In all that you do as women, do not forget that we are all children of God. If you get that doctrine in place, with that rule established, it will serve you well in times when you are confronted with those who equate humankind with animals.[27]

More recently, President Packer included anti-evolutionary statements in a Church-sanctioned article prepared for a special 2008 Ensign issue all about Jesus Christ. He wrote,

> We are taught in Genesis, in Moses, in Abraham, in the Book of Mormon, and in the endowment that man's mortal body was made in the image of God in a separate creation. Had the creation come in a different way, there could have been no Fall.
>
> If men were merely animals, then logic favors freedom without accountability.
>
> How well I know that among learned men are those who look down at animals and stones to find the origin of man. They do not look inside themselves to find the spirit there. They train themselves to measure things by time, by thousands and by millions, and say these animals called men all came by chance.[28]

In 2007, Elder Nelson reaffirmed his opposition to evolution during the following exchange with "The Pew Forum on Religion and Public Life," a major religious research organization:

> *Pew Forum*: Different denominations deal differently with questions about life's origins and development. Conservative denominations tend to have more trouble with Darwinian evolution. Does the Church have an official position on this topic?
>
> *Russell M. Nelson*: We believe that God is our Creator and that he has created other forms of life. It's interesting to me, drawing on my 40 years' experience as a medical doctor, how similar those species are. We developed open-heart surgery, for example, experimenting on lower animals simply because the same Creator made

> the human being. We owe a lot to those lower species. But to think that man evolved from one species to another is, to me, incomprehensible.
>
> *Pew Forum*: Why is that?
>
> *Russell M. Nelson*: Man has always been man. Dogs have always been dogs. Monkeys have always been monkeys. It's just the way genetics works.[29]

And Elder D. Todd Christofferson spoke out against evolution in the October 2010 General Conference. Commenting on the eroding effect it has upon moral values, he said:

> Those who believe that our bodies are nothing more than the result of evolutionary chance will feel no accountability to God or anyone else for what they do with or to their body. We who have a witness of the broader reality of premortal, mortal, and postmortal eternity, however, must acknowledge that we have a duty to God with respect to this crowning achievement of His physical creation.[30]

Finally, like so many other brethren, President Thomas S. Monson warned that scientific ideas that oppose revealed doctrine on the creation have the potential to harm us spiritually, if we allow those ideas to gain sway in our minds. In the following statement he counsels us to hold fast to God's word and to not let opposing ideas from science destroy our faith.

> Should doubt knock at your doorway, just say to those skeptical, disturbing, rebellious thoughts: "I propose to stay with my faith, with the faith

> of my people. I know that happiness and contentment are there, and I forbid you agnostic, doubting thoughts to destroy the house of my faith. I acknowledge that I do not understand the processes of creation, but I accept the fact of it. I grant that I cannot explain the miracles of the Bible, and I do not attempt to do so, but I accept God's word. I wasn't with Joseph, but I believe him. My faith did not come to me through science, and I will not permit so-called science to destroy it."[31]

Conclusion

The prophets will never lead us astray and the majority of the brethren will never be on the wrong side of a major issue. On the matter of evolution, our prophets, seers, and revelators have never openly proclaimed Darwin's theory to be true. Rather, on several occasions they have proclaimed macroevolutionary ideas to be false, both in public and Church venues.

Many of the statements presented in this chapter have come in General Conference addresses. The overall impression is that the teachings of the prophets and apostles are so overwhelmingly opposed to major aspects of evolution that no serious consideration can be given for sympathy toward the theory from God's servants. In their teachings we have the assurance that humankind did not evolve from lower orders of life.

Notes

[1]George Q. Morris, *General Conference*, October 1956.

[2]Delbert L. Stapley, *BYU Speeches*, May 5, 1964.

[3]Marion G. Romney, *General Conference*, April 1953.

[4]Marion G. Romney, *General Conference*, April 1959.

[5]Marion G. Romney, *General Conference*, April 1969.

[6]Marion G. Romney, "Man - A Child of God," *General Conference*, April 1973.

[7]Francis M. Gibbons, *Joseph Fielding Smith: Gospel Scholar, Prophet of God* (Salt Lake City: Deseret Book Co., 1992), 291.

[8]Joseph Fielding Smith, *Doctrines of Salvation*, 3 vols., ed. Bruce R. McConkie (Salt Lake City: Bookcraft, 1954-1956), 1:140. Emphasis in original.

[9]Joseph Fielding Smith, *Doctrines*, 1:141-142. Emphasis in original.

[10]Ezra Taft Benson, "God's Hand in Our Nation's History," *BYU Speeches*, Mar. 28, 1976.

[11]Bruce R. McConkie, *Mormon Doctrine*, 2nd Ed. (Salt Lake City: Deseret Book Co., 1979), 256. Emphasis in original.

[12]Bruce R. McConkie, *A New Witness for the Articles of Faith* (Salt Lake City: Deseret Book Co., 1985), 98–99.

[13]Bruce R. McConkie, "The Seven Deadly Heresies," *BYU Speeches*, June 1, 1980.

[14]Bruce R. McConkie, "The Foolishness of Teaching," Address given religious instructors, seminary and institute teachers in 1981.

[15]Bruce R. McConkie, *The Millennial Messiah: The Second Coming of the Son of Man* (Salt Lake City: Deseret Book Co., 1982), 424.

[16]Bruce R. McConkie, "Christ and the Creation," *Ensign*, June 1982.

[17]Mark E. Petersen, *Adam: Who Is He?* (Salt Lake City: Deseret Book Co., 1976), 39–51.

[18]Mark E. Petersen, *Moses: Man of Miracles* (Salt Lake City: Deseret Book Co., 1977), 159–166.

[19]Harold B. Lee, "Find the Answers in the Scriptures," *Ensign*, Dec. 1972.

[20]Ezra Taft Benson, "The Book of Mormon is the Word of God," *General Conference*, Apr. 1975. See also Ezra Taft Benson,*A Witness and a Warning: A Modern-Day Prophet Testifies of the Book of Mormon* (Salt Lake City: Deseret Book Co., 1988), 6.

[21]Ezra Taft Benson, *General Conference*, Apr. 1970. See also Ezra Taft Benson, *God, Family, Country: Our Three Great Loyalties* (Salt Lake City: Deseret Book Co., 1974), 225.

[22]Howard W. Hunter, *That We Might Have Joy*, (Salt Lake City: Deseret Book Company, 1994), 22.

[23]Gordon B. Hinckley, *Teachings of Gordon B. Hinckley* (Salt Lake City: Deseret Book., 1997), 298, emphasis added.

[24]Russell M. Nelson, *The Power within Us* (Salt Lake City: Deseret Book Co., 1988), 9.

[25]Boyd K. Packer, "The Mystery of Life," *General Conference*, October 1983.

[26]Boyd K. Packer, "The Pattern of Our Parentage," *General Conference*, October 1984.

[27]Boyd K. Packer, "Children of God," *BYU Women's Conference*, May 5, 2006.

[28]Boyd K. Packer, "Who Is Jesus Christ?" *Ensign*, March 2008.

[29]See "Pew Forum Interviews Church Leaders," LDS Newsroom, www.newsroom.lds.org/ldsnewsroom/eng/commentary/pew-forum-interviews-church-leaders.

[30]D. Todd Christofferson, "Reflections on a Consecrated Life," *General Conference*, October 2010.

[31]See Thomas S. Monson, "Decisions Determine Destiny," *BYU Speeches*, November 6, 2005; "Guideposts for Life's Journey," *BYU Speeches*, November 13, 2007; "Great Expectations," *BYU Speeches*, January 11, 2009.

Chapter 9

Scriptural Accounts of Adam and the Creation

Ryan Nilsson

You cannot be a faithful member and reject the scriptures – Bible, Book of Mormon, Doctrine and Covenants and Pearl of Great Price – for these are the standards of our faith. If you accept them you cannot accept organic evolution, for they are diametrically opposed. —Joseph Fielding Smith

Is evolution compatible with the gospel? Some would say yes. Many people who hold onto Darwin's theory of evolution and the gospel claim that they are compatible. Yet, as will be shown, when measured against scriptural accounts of Adam and the creation, there are several problems with this approach.

Because Darwinian and scriptural accounts of the creation differ so much, believers who accept both tend to uphold a different view of the scriptures that deal with the creation. To them a literal reading of the scriptures gives an inaccurate portrayal of events. When it comes to the creation, they believe that the scriptures are more allegorical than actual.

Why would some Latter-day Saints choose to compromise scriptural meaning in order to make room for evolution? There are probably many reasons. One may be that evolution is very popular and pervasive in our society. There are references to evolution in the news media, on the internet, television, radio, and especially academia. Given that Latter-day Saints are being bombarded with Darwinism, it is not surprising that some end up believing that it must be true. Still, others who are weak in testimony simply accept Darwin's full account without reservation.

Can one believe in both God and evolution? Certainly. But this chapter will show that one cannot reasonably believe in scriptural accounts of the creation and evolution. They present opposing views. There is no compromise that works. Either one accepts the truthfulness of the scriptures and rejects major aspects of evolutionary theory, or one accepts Darwinism and rejects fundamentals of the scriptures.

The Earth was a Paradise World

Evolution teaches that for billions of years the earth has for all intents and purposes been a fallen world, with organisms living and dying, and primal animal struggle being the rule (and the mechanism of progress). In contrast to this falsehood, we know that when the earth was originally created, it existed in a paradise state, free from this sort of corruption.

As the Tenth Article of Faith reads, "We believe . . . that the earth will be renewed and receive its paradisiacal glory [at the Second Coming of Christ]." That is, it was originally in a paradise state and it will again return to that state when the Lord returns. Before the Fall of Man, God declared his creation as "very good" (Moses 2:31), and had man not fallen, "all things which were created must have remained in the same state in which they were after they were created; and they must have remained forever, and had no end" (2 Nephi 2:22). That is, things would have remained indefinitely in the original paradisiacal state that the earth was created in, had the Fall of Adam not occurred. The theory of evolution allows no place for a paradisiacal earth.

Consider that if God initially created the world in a state of corruption, death, and turmoil—in other words, if God created a fallen world—then this would have made God the author of sin, evil, and corruption. Believers who accept this view are in the uncomfortable position of having to attribute sin and corruption directly to God!

The scriptures are clear that man brought sin and death into the world by means of transgression, and hence this is the reason for man's fallen nature and the earth's condition. For instance, the Book of Mormon teaches "there was no means to reclaim men from this fallen state, *which man had brought upon himself because of his own disobedience*" (Alma 42:12, emphasis added). The Book of Mormon also explains how God placed man upon the earth in a state where he was empowered to act for himself, and that through the use of this agency, mankind had precipitated the Fall (see 2 Ne. 2:14-23). Therefore, it is because of man's actions, and not God's, that we live in a fallen world.

Life Transplanted From Other Worlds

Darwinism teaches that life on this planet evolved over millions of years from simple, single-celled organisms into all the varieties of life today. The truth of the matter, however, is that many forms of life appeared relatively simultaneously, and that they appeared in a fully-developed condition. The reason for this is that life on our planet did not originate here; it was brought here from other worlds which God had already created. The scriptures tell us that in the beginning, seeds of all kinds were placed in the earth. These seeds gave rise to the first vegetation.

The scriptures say: "And the Gods organized the earth to bring forth grass from its own seed, and the herb to bring forth herb *from its own seed*, yielding seed after his kind; and the earth to bring forth the tree *from its own seed*, yielding fruit, *whose seed could only bring forth the same in itself, after his kind*" (Abraham 4:12, emphasis added). This scripture suggests that the first seeds came from other worlds; they did not evolve.

Prophets have taught the concept of seeds coming from other worlds. Speaking on this matter, Joseph Smith asked rhetorically, "Whenever did a tree or anything spring into existence without a progenitor?"[1] And President Brigham Young taught that "[Adam] was the person who *brought the animals and the seeds from other planets to this world*, and brought a wife with him and stayed here."[2]

President Joseph Fielding Smith concurred with the view that earth life originated on another planet when he wrote:

> does it not appear to you that it is a foolish and ridiculous notion that when God created this earth he had to begin with a speck of protoplasm, and take millions of years, if not billions, to bring conditions to pass by which his sons and daughters might obtain bodies made in his image? Why not the shorter route and transplant them from another earth as we are taught in the scriptures?"[3]

President Smith makes a valid point. Why would God

bother with evolution when He could simply take existing life from other planets to populate our own world? In any case, life on our planet was transplanted here from other worlds. It did not "evolve" into existence as evolutionists claim.

All Animals and Things Created Spiritually First

There is little, if any room at all, for the spiritual or metaphysical within the theory of evolution. Evolution is wholly materialistic and does not acknowledge the existence of the spirit. It is significant therefore that the scriptures teach us that God first fashioned living things in the spirit before creating them in physical form. He declared, "For I, the Lord God, created all things, of which I have spoken, spiritually, before they were naturally upon the face of the earth" (Moses 3:5).

The concept of a spiritual creation does not fit well within the context of evolution. For example, did spiritual life forms evolve? If there was a physical evolution, then it would seem that a spiritual evolution would have been necessary for the spiritual forms to correspond with their destined physical forms. This being the case, how would that work? Evolution is supposed to have been in response to the primal struggle involving suffering and death – the survival of the fittest – conditions that simply did not exist in the premortal realms. In fact, it was the very absence of such brutal conditions in the premortal realms that necessitated a mortal experience on earth for our testing and trial in the first place.

And what about vestigial structures? If there are useless structures in our bodies, does that mean that those structures are also in our spirits? It would seem so. The scriptures tell us that "that which is spiritual [is] in the likeness of that which is temporal; and that which is temporal in the likeness of that which is spiritual; the spirit of man in the likeness of his person, as also the spirit of the beast, and every other creature which God has created" (Doctrine and Covenants 77:2). If we accept both this scripture and evolution, that would suggest that there are countless transitional spiritual forms and spirits with intentionally useless vestigial structures! These ideas don't make much sense, especially in the context of eternity. The doctrine of the spiritual creation makes mixing evolution and the gospel more problematic than most people probably realize.

Man Was Created in the Image of God from the Beginning

Evolution teaches that man did not originate as human beings in the beginning, but instead started out as a single-celled organism, and then slowly, through a great many successive generations, evolved to his present state. In contrast to this, the scriptures teach that *in the beginning*, man was fashioned *in the image of God*: "So God created man in his own image, in the image of God created he him; male and female created he them" (Genesis 1:27; Moses 2:27; Abraham 4:27).

Referring to the Savior, the scriptures also declare,

> he should take upon him the image of man, and *it should be the image after which man was created in the beginning*; or in other words, he said that man was created after the image of God. (Mosiah 7:27, emphasis added)

> [M]an *in the beginning* was created *after the image of God.* (Alma 18:34, emphasis added)

> Seest thou that ye are created after mine own image? Yea, *even all men were created in the beginning after mine own image.* (Ether 3:15, emphasis added)

The following statement from the 1909 First Presidency sums up these teachings: "It was shown to the brother of Jared that all men were created in the beginning after the image of God; whether we take this to mean the spirit or the body, or both, it commits us to the same conclusion: Man began life as a human being, in the likeness of our Heavenly Father."[4]

Adam was a Historical Person

The theory of evolution does not acknowledge Adam. For most evolutionists, scriptural accounts of the creation are mythical stories. In fact, the very existence of Adam is necessarily challenged by evolution theory, especially as the story of Adam is presented in the scriptures. But for Latter-day Saints who accept the scriptures, Adam was a real, historical person. He was the first man and the progenitor of all mankind. The scriptures indicate this much. In Moses 6:9 we read that when God created Adam and Eve, they were "living souls." How could they be called living souls if they were not real people?

The revelations of Joseph Smith chronicle the lineage of the priesthood, and trace it directly back to Adam. The Doctrine and Covenants say,

> This order was instituted in the days of Adam, and came down by lineage in the following manner: From Adam to Seth, who was ordained by Adam at the age of sixty-nine years, and was blessed by him three years previous to

> his (Adam's) death. (Doctrine and Covenants 107:41–42)

Consider that if Adam was not a real person there would be no reason to chronicle the descent of the priesthood through his lineage. There would be no legitimate reason for the Lord to establish this connection of priesthood descent back to him; it would be an utterly useless exercise at best, and certainly misleading.

It is significant that Joseph Smith made frequent reference to Adam, describing him as intelligent and mighty. The prophet taught "That [Adam] received revelations, commandments and ordinances at the beginning" and that this reality "is beyond the power of controversy."[5] The prophet also revealed that Adam is none other than the Archangel Michael, who stands next to Christ himself in authority as the next "Great High Priest" to whom all subsequent prophets must give an account of their stewardship.[6] Does this sound like a primitive man?

Joseph Smith frequently talked about events in the creation narrative, including our first parents in the Garden of Eden, in a literal way. He even went as far as to identify the precise location of the Garden of Eden in Jackson County, Missouri.[7] He recalled, "I saw Adam in the valley of Adam-ondi-Ahman."[8] The prophet even heard the voice of Michael, who is Adam. He related, "And again, what do we hear? . . . The voice of Michael on the banks of the Susquehanna, detecting the devil when he appeared as an angel of light!" (Doctrine and Covenants 128:20). And Adam appeared to the Prophet Joseph Smith along with Gabriel, Raphael, and "divers angels . . . all declaring their dispensation, their rights, their keys, their honors, their majesty and glory, and the power of their priesthood" (Doctrine and Covenants 128:21). Does this sound like Adam is a mere allegory or a myth?

Indeed, Joseph Smith was a treasure trove of heavenly knowledge on Adam. He knew the truth and he assured us that he knew it when he said, "I could go back and trace every subject of interest concerning the relationship of God to man, if I had time. I can enter into the mysteries; I can enter largely into the eternal worlds."[9] His knowledge on the matters of creation and eternity was truly extensive, far beyond the knowledge of any other single human being in our modern era of history.

Would Joseph Smith have changed his opinions about Adam had he been contemporary with Darwin and had learned about evolution? It is unthinkable and presumptuous to suggest that Darwin, or any modern-day evolutionist, might have anything meaningful to add to the prophet's understanding about our father Adam, of whom he had first-hand knowledge.

Joseph Smith said of his understanding, "I am learned, and know more than all the world put together. The Holy Ghost does, anyhow, and he is within me, and comprehends more than all the world: and I will associate myself with Him."[10] Joseph Smith the Prophet also declared, "I know the scriptures and understand them,"[11] and he stated unequivocally, "I believe in the fall of man, as recorded in the Bible."[12]

The evidence from the scriptures and the teachings of the Prophet Joseph Smith are overwhelming. If we accept the scriptures and the prophetic calling of Joseph Smith, then we are completely bound to believing that Adam was a real, historical person. It therefore follows that if Adam was not a real person, as Darwinism teaches, then our religion is false.

Adam was the First Man

Bringing Darwin's ideas into harmony with the gospel requires adaptations to the origin of Adam, or to Darwin's ideas. Some theistic evolutionists assume that Adam was one person among many who descended from hominid-like creatures. This would require Adam having human parents and ancestors who were living and dying on the earth—an evolutionary leap from hominid to the first man Adam in one short generation would not fit too well with the theory of evolution. Thus according to this view, Adam was not the first man, yet the scriptures repeatedly declare that Adam was the first man. A small sampling should suffice to illustrate this point:

> And the first man of all men have I [i.e. God] called Adam. (Moses 1:34) And I, the Lord God, formed man from the dust of the ground, and breathed into his nostrils the breath of life; *and man became a living soul, the first flesh upon the earth, the first man also.* (Moses 3:7, emphasis added)
>
> [The Priesthood] came down from the fathers, from the beginning of time, yea, even from the beginning, or before the foundation of the earth, down to the present time, even the right of the firstborn, or *the first man, who is Adam, or first father, through the fathers unto me.* (Abraham 1:3, emphasis added)
>
> [Abel] received the priesthood by the commandments of God, by the hand of *his father Adam, who was the first man.* (Doctrine and Covenants 84:16, emphasis added)
>
> And to bring about his eternal purposes in the

> end of man, after *he had created our first parents*, and the beasts of the field and the fowls of the air, and in fine, all things which are created. (2 Nephi 2:15, emphasis added)

In addition to these declarations there are several scriptures which tell us that all of the children born in this world are descendants of Adam and Eve. The scriptures declare, "And Adam called his wife's name Eve, *because she was the mother of all living*; for thus have I, the Lord God, called the *first of all women*" (Moses 4:26, emphasis added; see also Genesis 3:20). And Lehi taught, "And they [Adam and Eve] have brought forth children; yea, *even the family of all the earth*" (2 Nephi 2:20, emphasis added).

The scriptures elsewhere indicate: "And [God] hath made *of one blood all nations of men for to dwell on all the face of the earth*, and hath determined the times before appointed, and the bounds of their habitation" (Acts 17:26, emphasis added). These teachings oppose the idea that humans were living on the earth longer than 6,000 years ago, and that Adam's family was only one among many.

Adam, a Son of Deity

Evolution teaches that mankind is descended from animals, and that we evolved from the lower orders of creation. But the gospel teaches that mankind is no less than the literal offspring of deity. On Mars' Hill the Apostle Paul taught that "we are the offspring of God" (Acts 17:28–29).

How was the creation of Adam carried out? The scriptures tell us that Adam was "born into the world by water, and blood, and the spirit [i.e. the natural elements of birth]" (Moses 6:59, emphasis added). They also assure us that Adam "was the son of God, with whom God, himself, conversed" (Moses 6:22), and that Adam was the "firstborn" into the world (Abraham 1:3). And so the physical bodies of Adam and Eve were literally begotten by heavenly beings. It is true that the scriptures say that man was "formed . . . from the dust of the ground" (Moses 3:7; Abraham 5:7; Genesis 2:7); however, this is a cryptic verse in scripture that we are told elsewhere in holy writ means simply to be born: "inasmuch as ye [Adam] were *born into the world* by water, and blood, and the spirit, which I have made, *and so became of dust a living soul*" (Moses 6:59, emphasis added).

If Adam is a literal son of deity, then this excludes the possibility of him being the offspring of another mortal, hominid, or any sort of animal. The true understanding of man's origin is exactly opposite to the demeaning implications of Darwinism. The real heritage of mankind is exalted parentage, not a swirling primordial pond.

In 1909, the First Presidency reaffirmed that deity is man's lineage. In an official statement concerning evolution, they said, "The Church of Jesus Christ of Latter-day Saints, basing its belief on divine revelation, ancient and modern, proclaims man to be the direct and lineal offspring of Deity."[13] Elder James E. Talmage agreed when he testified that "[Man] is born in the lineage of Deity, not in the posterity of the brute creation."[14]

Adam Lived 6,000 Years Ago, Circa 4000 B.C.

Evolution teaches that the primitive predecessors of mankind first emerged around two million years ago and human beings as we know them appeared several hundred thousand years ago. However, when this idea is measured against the scriptures, it is found wanting. We know from adding up the ages of the patriarchs over the time span of the history of the world that Adam lived about 6,000 years ago, circa 4000 B.C. We do not have to look exclusively to the Old Testament for this information; the ages of these patriarchs are corroborated by the revelations of Joseph Smith.

The revelations of Joseph Smith are found in the Book of Moses, chapter 6, verses 10–25 (Joseph Smith's inspired version of the Genesis account), the Doctrine and Covenants (sections 84 and 107 give the lineage of the priesthood and the ages at which the patriarchs were ordained), and in the Lectures on Faith, co-authored by the prophet (in Lecture Second is given an account of the ages of the patriarchs). In addition to this, the Prophet Joseph Smith confirmed the timeframe of human history when he remarked, "The world has had a fair trial for six thousand years; the Lord will try the seventh thousand Himself."[15]

Fall of Adam was Necessary for the Atonement

Another problem with harmonizing evolution and the gospel deals with the Fall of Adam. We know from the scriptures that the Fall is the necessary condition for the Atonement of Jesus Christ. The Fall of Adam brought sin and death into the world, and the Atonement and Resurrection of Jesus Christ reconciled those two problems. In other words, according to the scriptures the entire reason Jesus Christ came into the world to save mankind was to remedy the condition brought onto the world by the Fall of Adam. If evolution is true and there was no Fall of Man, there would be no need for an Atonement. Thus evolution unwittingly undermines the most central and important concept of our religion, that of having a Christ.

Consider the following scriptures highlighting the inseparable link between the Fall and the Atonement:

> For as death hath passed upon all men, to fulfill the merciful plan of the great Creator, there must needs be a power of resurrection, and *the resurrection must needs come unto man by reason of the fall*; and the fall came by reason of transgression. (2 Nephi 9:6, emphasis added)
>
> Behold, [God] created Adam, and by Adam came the fall of man. And *because of the fall of man came Jesus Christ*, even the Father and the Son; and because of Jesus Christ came the redemption of man. (Mormon 9:12, emphasis added)
>
> [A]s in Adam, or by nature, they fall, even so the blood of Christ atoneth for their sins. (Mosiah 3:16)
>
> For since by man came death, by man came also the resurrection of the dead. For as in Adam all die, even so in Christ shall all be made alive. (1 Corinthians 15:21–22)

The scriptures are clear: If there was no Fall there could be no Atonement. Belief in Adam and his Fall is essential to believing in Christ. This is a major reason why belief in evolution theory is so incredibly dangerous; it is spiritually perilous precisely because it undermines the logical foundation for the role of Jesus Christ in the Plan of Salvation.

Do those who wish to accept both evolution and the scriptures consider this grave problem? And are they able to circumvent the inseparable relationship between the Fall and the Atonement shown to be necessary within the scriptures? They cannot! It is a hopeless venture. But the truth, instead, has lain right before us in the scriptures all along. All that is needed is for us to exercise the requisite faith to believe it.

Notes

[1]Joseph Smith, *Teachings of the Prophet Joseph Smith*, ed. Joseph Fielding Smith (Salt Lake City: Deseret Book Company, 1976), 373.

[2]Brigham Young, *Journal of Discourses*, ed. George D. Watt, et al. 26 vols. (Liverpool: F. D. Richards, et al., 1854–1886), 3:319, emphasis added.

[3]Joseph Fielding Smith, *Man: His Origin and Destiny*, (Salt Lake City: Deseret Book Company, 1965), 277.

[4]Joseph F. Smith, John R. Winder, and Anthon H. Lund, "The Origin of Man," *Improvement Era*, 13 (November 1909), 75–81.

[5]Joseph Smith, *Teachings of the Prophet Joseph Smith*, ed. Joseph Fielding Smith (Salt Lake City: Deseret Book Company, 1976), 168.

[6]Joseph Smith, *Teachings*, 158. See also D&C 27:11.

[7]Joseph Smith, *Encyclopedia of Joseph Smith's Teachings*, ed. Dahl & Cannon (Salt Lake City: Deseret Book Company, 1997), 277.

[8]Joseph Smith, *Teachings of the Prophet Joseph Smith*, ed. Joseph Fielding Smith (Salt Lake City: Deseret Book Company, 1976), 158.

[9]Joseph Smith, *Teachings*, 359.

[10]Joseph Smith, *Teachings*, 350

[11]Joseph Smith, *Teachings*, 357.

[12]Joseph Smith, *Encyclopedia of Joseph Smith's Teachings*, ed. Dahl & Cannon (Salt Lake City: Deseret Book Company, 1997), 238.

[13]See Joseph F. Smith, John R. Winder, and Anthon H. Lund, "The Origin of Man," *Improvement Era*, 13 (November 1909), 75–81.

[14]James E. Talmage, "The Earth and Man," *Deseret News*, November 21, 1931, 11–21.

[15]Joseph Smith,*Teachings of the Prophet Joseph Smith*, ed. Joseph Fielding Smith (Salt Lake City: Deseret Book Company, 1976), 252.

Chapter 10

Incompatibility of Evolutionary Concepts and Gospel Teachings

Ryan Nilsson

No man can serve two masters: for either he will hate the one, and love the other; or else he will hold to the one, and despise the other – Matthew 6:24

Catastrophism vs. Uniformitarianism

Basic to the theory of evolution is the assumption of uniformitarianism. This is the idea that the earth we see today is the product of slow, gradual processes. This belief largely denies catastrophic events as explanations for many of the different geological curiosities on the earth. However, the scriptures call this idea into serious question. A massive catastrophic event has happened—the Flood of Noah.

The Apostle Peter prophesied that there would be many in the last days that would assert a false uniformitarian theory and thereby deny the Flood and other truths of the gospel. He wrote:

> Knowing this first, that there shall come in the last days scoffers, walking after their own lusts, And saying, Where is the promise of his [Christ's] coming? for since the fathers fell asleep, *all things continue as they were from the beginning of the creation.* For this they willingly are ignorant of, that by the word of God the heavens were of old, and the earth standing out of the water and in the water: Whereby *the world that then was, being overflowed with water, perished*: But the heavens and the earth, which are now, by the same word are kept in store, reserved unto fire against the day of judgment and perdition of ungodly men. (2 Peter 3:3–7, emphasis added)

This scripture tells us that the Flood, that great catastrophic deluge, completely destroyed the previous world or the "world that then was."

We also know that the earth has undergone many changes since the creation. The calamitous destruction prior to the visit of Christ to the Nephites is another example: "for behold, the whole face of the land was changed, because of the tempest and the whirlwinds, and the thunderings and the lightnings, and the exceedingly great quaking of the whole earth" (3 Nephi 8:12). Therefore, keeping the scriptures in mind it seems prudent for us to re-evaluate Darwinism's claim that all living things on earth evolved through slow and gradual processes. Catastrophic events like the Flood change our world in significant ways.

Longevity of the Patriarchs

From an evolutionary perspective, it seems impossible that the life span of human beings in the distant past could reach far beyond our present day's average. And yet the scriptures give accounts of pre-Flood men with lifespans reaching beyond 900 years! For instance, the Book of Genesis gives an account of the amazing longevity of patriarchs with specific details as to their ages (see Genesis 5:5–32; 9:28–29). Here are some lifespans of the early patriarchs:

Patriarch (Age at death in years)

Adam (930)

Seth (912)

Enos (905)

Cainan (910)

Mahalaleel (895)

Jared (962)

Enoch (translated at age 430 - see Moses 8:1)

Methuselah (969)

Lamech (777)

Noah (950)

In addition to this, when the generations of men that had lived close to the time of the Flood were drawing to a close, the Prophet Jacob told the Pharaoh of Egypt, "The days of the years of my pilgrimage are an hundred and thirty years: few and evil have the days of the years of my life been, and have not attained unto the days of the years of the life of my fathers in the days of their pilgrimage" (Genesis 47:9). In other words, Jacob testified to Pharaoh that, although he was 130 years old, his life span was miniscule compared to the fathers before him.

These long life spans are inexplicable within the theory of evolution, and yet other scriptures confirm the reality of the Old Testament account. The Book of Moses (see Moses 6:12–27), the Doctrine and Covenants (see D&C 107:41–52), and the *Lectures on Faith* (Lecture 2:37–43) corroborate the great ages of the patriarchs.

No Death Before the Fall

Absolutely crucial to the theory of evolution is the presence of death in this world for eons of time. According to evolution, death is how the human race eventually came to be, having arisen from lower animals over successive generations. Thus human beings are believed to have been living and dying for thousands of years prior to the time of Adam. The scriptures clearly contradict this idea. They state that *death did not bring man into the world, rather man brought death into the world.* The following examples indicate that humans were not dying on the earth before the Fall of Adam.

> And [Enoch] said unto them: Because that Adam fell, we are; and *by his fall came death*; and we are made partakers of misery and woe. (Moses 6:48, emphasis added)

> Wherefore, as *by one man sin entered into the world, and death by sin*; and so death passed upon all men. . . . [B]y one man's offence death reigned. (Romans 5:12, 17, emphasis added)

> [B]y reason of transgression cometh the fall, *which fall bringeth death.* (Moses 6:59, emphasis added)

> And now, behold, if Adam had not transgressed he would not have fallen, but he would have remained in the garden of Eden. And all things which were created must have remained in the same state in which they were after they were created; and they must have remained forever, and had no end. (2 Nephi 2:22)

> [T]he fall had brought upon all mankind a spiritual death *as well as a temporal.* (Alma 42:9)

> For since by man [Adam] came death. (1 Corinthians 15:21)

These declarations, that there was no death before the Fall and that Adam's transgression in the Garden of Eden brought death into the world, are too numerous to be merely discarded in favor of a theistic evolutionary account of the creation.

No Vestigial Structures

Evolution teaches that since man has evolved from lower forms of life, he has within himself many vestigial or unnecessary structures. Vestigial structures are thought to be leftover relics from our evolutionary heritage. Commonly the appendix and tail bone are cited as vestigial structures. However, the idea that man has useless structures is contrary to teachings in the scriptures.

The scriptures say the following about the human body: "let not the head say unto the feet it hath no need of the feet; for without the feet how shall the body be able to stand? Also the body *hath need of every member* . . . that the system may be kept perfect" (Doctrine and Covenants 84:109–110, emphasis added. See also 1 Corinthians 12:14–23). Likewise Joseph Smith taught that the

Priesthood "may be illustrated by the figure of the human body, which has different members, which have different offices to perform; *all are necessary in their place, and the body is not complete without all the members.*"[1] Thus according to the scriptures, there are no useless structures. Every structure of the body has a divinely appointed use and purpose.

Proper Orders and Natural Forms

Essential to the theory of evolution is the idea that all creatures are related because they share a common ancestor. Diversification is believed to have occurred when organisms gradually changed over millions of years, evolving from one kind of animal into another. This process of gradual change across life forms would require significant genetic variations, yet the word of God (and indeed nature itself) indicates that genetic variations capable of changing one kind of creature into another do not occur.

The scriptures tell us that living things have been commanded to bring forth "after their kind" (see, for example, Moses 2:21, 25). The idea that life is assigned to separate classes or kinds is repeated so much in the creation narratives that it makes little sense to suggest otherwise. The Apostle Paul taught this principle when he wrote to the Corinthians that "All flesh is not the same flesh: but there is one kind of flesh of men, another flesh of beasts, another of fishes, and another of birds" (1 Corinthians 15:39).

Living things cannot go beyond the bounds which God has assigned. The scriptures say that the Lord "organized the earth to bring forth grass from its own seed, and the herb to bring forth herb from its own seed, yielding seed after his kind; and the earth to bring forth the tree from its own seed, yielding fruit, whose seed could only bring forth the same in itself, after his kind" (Abraham 4:12). Joseph Smith agreed. He taught that it was impossible for life to bring forth in any other way than the pattern of its parentage. He affirmed that "it is a decree of the Lord that every tree, plant, and herb bearing seed should bring forth of its kind, and cannot come forth after any other law or principle."[2]

The scriptures also suggest that living things do not go beyond their bounds and evolve into different kinds of life. God separated all forms of life into great classes with unsurpassable boundaries. Granted, there may be fabulous variety within these classes, but these classes are fixed by heavenly decree and cannot be transcended. The same law exists in other worlds created by the Lord where animals are segregated into "classes of beings in their destined order or sphere of creation" (Doctrine and Covenants 77:3). All animals throughout time and eternity are subject to genetic limits and bounds, and are restricted to their particular spheres of existence. They cannot transition into different kinds of creatures and exceed the bounds which God has decreed.

Lastly, regarding the resurrection, the scriptures teach, "Therefore, all things shall be restored to their proper order, everything to its natural frame—mortality raised to immortality" (Alma 41:4). According to this verse, every living creature will be restored to its original state, or its God-appointed natural frame with the exception of deformities and unintended changes to natural form.

Yet within evolution theory there are no deviant vestigial structures or changes to correct in the resurrection. The reason for this is that, since there is no designed form, those vestigial structures are an integral part of a species' existence. Physical change is constant and aimless in evolution, and there can be no "proper order" or "natural frame" to which a creature should conform other than the wasteful mass of vestigial structures and partly-formed organs provided haphazardly by evolution. There would have been countless transitions in the lineage of living creatures as they evolved from one type of creature into another. There can be no correct form or proper frame to discern among these; it is just a continuous string of perpetual change.

If this position were true, then what is the correct form to which a newly evolved creature should be resurrected? Should it be resurrected to the natural form intended by the Lord or some other evolved form? Thus we see that for theistic evolutionists who choose to incorporate the reality of the resurrection within Darwinism, there are considerable challenges.

The Flood – A Literal, Historical Event

Many evolutionists deny that there was ever a catastrophic, worldwide flood. Elder Mark E. Petersen expressed that, "Some doubt that there was a flood, but by modern revelation we know that it did take place."[3] Howard W. Hunter confirmed this sentiment as well.[4] Evolutionists deny the Flood because it is opposes their view of slow and gradual change (uniformitarianism). The scriptures, however, proclaim Noah's Flood as a literal, historical event.

The reality of the Flood is avowed in the teachings of Jesus who likened the days of the Flood to the days of His Second Coming. He said,

> But as the days of Noe were, so shall also the coming of the Son of man be. For as in the days that were before the flood they were eating and drinking, marrying and giving in marriage, until the day that Noe entered into the ark, And knew not until the flood came, and took them

> all away; so shall also the coming of the Son of man be. (Matthew 24:37–39)

If the Flood did not actually occur, then the effectiveness of Jesus' use of it as a comparison to his Second Coming is severely diminished. Where is the convincing power in the Savior's statement if the Flood did not happen? Clearly the second part about the Second Coming in the foregoing statement is dependent upon the first portion, the Flood. Because Jesus referred to both events in a literal sense, rejecting the Flood compels one to also reject the Second Coming.

Joseph Smith similarly taught that the Flood is analogous to the Second Coming. He taught that "In the days of Noah, God destroyed the world by a flood, and He has promised to destroy it by fire in the last days."[5] If one is true then so is the other. If one is not true then neither is the other. It is inconsistent to say that the destructive Flood chronicled in the scriptures is fictional while saying that the destruction that will occur during the Second Coming is factual.

The Prophet Joseph Smith also declared that "Noah, who is Gabriel . . . stands next in authority to Adam in the Priesthood; he was called of God to this office, and was the father of all living in this day, and to him was given the dominion."[6] Gabriel (Noah) was the "father of all living" in his day because "in the days of Noah, while the ark was a preparing, wherein few, that is, eight souls were saved by water" (1 Peter 3:2).

Thus Noah and his posterity were obliged to repopulate the earth much the same way Adam and Eve were obliged to populate it the first time. The reality of Noah's Flood is further established by President Joseph F. Smith's grand vision on the Redemption of the Dead. In that vision, President Smith reported seeing many of the "great and mighty ones" whom God had revealed himself to throughout time. Among these were none other than "Noah, who gave warning of the flood" (Doctrine and Covenants 138:41). This constitutes incontrovertible evidence from modern scripture regarding the reality of Noah and his warning of the great Flood.

The Book of Moses in the Pearl of Great Price also corroborates the reality of Noah's Flood. It says:

> And Enoch also saw Noah, and his family; that the posterity of all the sons of Noah should be saved with a temporal salvation; Wherefore Enoch saw that Noah built an ark; and that the Lord smiled upon it, and held it in his own hand; but upon the residue of the wicked the floods came and swallowed them up. (Moses 7:42–43)

The conclusion is fairly obvious. The scriptures and the teachings of the prophets indicate that Noah was a real person who warned of a Flood that actually happened. In contrast to the evolutionary assumption of uniformitarianism, the Flood was a catastrophic event that brought about rapid and dramatic changes to the earth.

If evolutionists accept the notion of a flood at all, they claim that it may have happened on a limited scale, such as in a localized region of Mesopotamia. But the scriptures maintain that a Flood did indeed cover the entire earth. They say:

> Fifteen cubits upward did the waters prevail; and the mountains were covered. And all flesh died that moved upon the earth, both of fowl, and of cattle, and of beast, and of every creeping thing that creepeth upon the earth, and every man: All in whose nostrils was the breath of

> life, of all that was in the dry land, died. And every living substance was destroyed which was upon the face of the ground, both man, and cattle, and the creeping things, and the fowl of the heaven; and they were destroyed from the earth: and Noah only remained alive, and they that were with him in the ark. (Genesis 7:20–23)

The Book of Mormon, which Joseph Smith declared to be "the most correct of any book on earth, and the keystone of our religion,"[7] corroborates the Bible's teaching that the Flood was a worldwide catastrophic event. The Prophet Ether in the Book of Mormon taught that "after the waters [of the Flood] had receded from off the face of this land [i.e. the American continent] it became a choice land above all other lands, a chosen land of the Lord" (Ether 13:2). Therefore the Flood waters had covered the land of the Americas, as well as everywhere else in the world because the Ark came to rest "upon the mountains of Ararat" (Genesis 8:4), which is in modern-day Turkey.

Further evidence of a worldwide flood is found in teachings on the earth's baptism. The scriptures tell us that "the earth abideth the law of a celestial kingdom, for it filleth the measure of its creation, and transgresseth not the law" (Doctrine and Covenants 88:25). Part of the law of the celestial kingdom, and indeed certainly its most conspicuous requirement, is the law of baptism. All accountable people are subject to the requirement of baptism if they wish to inherit the celestial kingdom (Doctrine and Covenants 76:51). The earth itself, as an actual sentient being aware of concepts such as sin and righteousness, is no different in this requirement. As the scriptures relate, the earth itself spoke with a voice to Enoch and lamented,

> Wo, wo is me, the mother of men; I am pained,

> I am weary, because of the wickedness of my children. When shall I rest, and be cleansed from the filthiness which is gone forth out of me? When will my Creator sanctify me, that I may rest, and righteousness for a season abide upon my face? (Moses 7:48)

The prophets have testified of the earth's requirement to be baptized. For example, commenting on the Flood story, President John Taylor taught that "the earth was immersed" and that "It was a period of baptism."[8] President Brigham Young similarly taught:

> The earth, in its present condition and situation, is not a fit habitation for the sanctified; but it abides the law of its creation, has been baptized with water, will be baptized with fire and the Holy Ghost, and by-and-by will be prepared for the faithful to dwell upon.[9]

The Spiritually Destructive Effect of Evolution Theory

Evolution teaches that mankind is just another animal, simply a higher-evolved beast. The gospel, on the other hand, teaches that we are children of God. The difference between these two worldviews can have a tremendous effect upon a person's moral compass. The evolutionary doctrine which teaches that we are merely animals leads people to disregard codes of moral conduct. The reasoning goes that since beasts are not subject to moral laws, and man is a beast, then man is not subject to a higher moral code. If man truly is a beast then this conclusion would be inescapable. This is the claim that Korihor, the antichrist from the Book of Mormon, used. He reasoned,

> that there could be no atonement made for the sins of men, but every man fared in this life according to the management of the creature; therefore every man prospered according to his genius, and that every man conquered according to his strength; and whatsoever a man did was no crime. (Alma 30:17)

Of course, these are the same conclusions drawn from evolution, if one takes it to its logical end. Abandoning moral responsibility to a higher power leads to sinful behavior. The scriptures indicate that this happened to the people who accepted Korihor's teachings: "And thus [Korihor] did preach unto them, leading away the hearts of many, causing them to lift up their heads in their wickedness, yea, leading away many women, and also men, to commit whoredoms—telling them that when a man was dead, that was the end thereof" (Alma 30:18).

The effect of such teachings is the same in every age of the world, including today. At various times in recent history, evolution has been used as a justification for people's diminishing moral compass. It is of significant note that evolution has served as the philosophical underpinning to communism, socialism, and Nazi fascism, as well as initiatives such as eugenics, genocide, and population control.[10] As a case in point, take Adolf Eichmann, a man known as "Hitler's Hangman." Eichmann specifically cited evolution as a justification for his murderous deeds during World War II. At his trial for war crimes he declared, "God's method of creation was to wipe out the handicapped and to wipe out the less fitted. And as the Jews are less fitted than our people, I have only helped God in his methods. I have only catalyzed God's way of working. And when I meet God I shall tell him so."[11]

Evolution has also been used to justify behavioral vices

such as promiscuity, teen pregnancy, abortion, violence, and uncivilized behavior. For example, MIT professor Steven Pinker explained infanticide—the killing of children by the parents—as an expression of evolutionary tendencies.[12] The implications of evolution theory—that man is a beast, and that he is not subject to God's moral code—can have a great effect on people's minds.

The notion that people are simply animals does not entice them to believe in God or to do good, but rather entices them to do whatever their carnal desires prompt them to do, in the same way that an animal is driven by carnal desires. This concept is hardly edifying.

Thankfully the premise that man is merely an advanced beast is incorrect. The gospel teaches us that we are children of God. As His children we are prompted by the Spirit to live His moral code. Moroni taught this doctrine in the following way:

> But behold, that which is of God inviteth and enticeth to do good continually; wherefore, every thing which inviteth and enticeth to do good, and to love God, and to serve him, is inspired of God. . . . [F]or every thing which inviteth to do good, and to persuade to believe in Christ, is sent forth by the power and gift of Christ; wherefore ye may know with a perfect knowledge it is of God. But whatsoever thing persuadeth men to do evil, and believe not in Christ, and deny him, and serve not God, then ye may know with a perfect knowledge it is of the devil. (Moroni 7:13, 16–17)

The true doctrine that men and women are children of God is edifying, and leads them to better behavior.

Conclusion

It is unreasonable to accept both evolution's theory of common descent and scriptural accounts of Adam and the creation because they are opposed to each other. If the scriptures are true, then many facets of evolution are false. By the same token, if evolution is true, then many aspects of the scriptures are false and untrustworthy. Both cannot be true.

The Savior taught, "No man can serve two masters: for either he will hate the one, and love the other; or else he will hold to the one, and despise the other" (Matthew 6:24). Still some persist in putting the theories of man ahead of the scriptures and suppose the latter to be mere allegorical expressions when they contradict manmade ideas.

The scriptures are our primary link with heaven, our handbook for understanding the doctrines of the kingdom. With the scriptures as our guide to true doctrine, we ought to accept their account of the creation and turn away from conflicting manmade theories. Perhaps President Joseph Fielding Smith said it best when he wrote,

> You cannot be a faithful member and reject the scriptures – Bible, Book of Mormon, Doctrine and Covenants and Pearl of Great Price – for these are the standards of our faith. If you accept them you *cannot* accept organic evolution, for they are diametrically opposed.[13]

Notes

[1]Joseph Smith, *Teachings of the Prophet Joseph Smith*, ed. Joseph Fielding Smith (Salt Lake City: Deseret Book Company, 1976), 112, emphasis added.

[2]Joseph Smith, *Teachings*, 197–198.

[3]Mark E. Petersen, "Follow the Prophets," *General Conference*, Oct. 1981.

[4]Howard W. Hunter, "Where, Then, Is Hope?" *General Conference*, Oct. 1970.

[5]Joseph Smith, *Encyclopedia of Joseph Smith's Teachings*, ed. Dahl & Cannon (Salt Lake City: Deseret Book Company, 1997), 204.

[6]Joseph Smith, *Teachings*, p.157.

[7]Joseph Smith, *Teachings*, p.194.

[8]John Taylor, *Journal of Discourses*, ed. George D. Watt, et al. 26 vols. (Liverpool: F. D. Richards, et al., 1854–1886), 26:75.

[9]Brigham Young as quoted in Joseph Fielding Smith, *Answers to Gospel Questions* (Salt Lake City: Deseret Book Company, 1963), 4:20.

[10]Socialism, of which communism and fascism are variations, is no doubt the most murderous ideology this world has ever seen, being responsible for the murder of more than 100 million people. Essentially its underlying philosophy is evolution (See "Marxism, Law and Evolution: Marxist law in both theory and practice," Creation Ministries International, accessed October 7, 2014, http://creation.com/marxism-law-and-evolution).

[11]See Marvin L. Lubenow, *Bones of Contention: A Creationist Assessment of Human Fossils*, Revised and Expanded Edition (Grand Rapids: Baker Books, 2004), 153.

[12]"Steven Pinker's Evolutionary 'Explanation' of Infanticide," Christian Research Institute, http://www.equip.org/articles/steven-pinkers-evolutionary-explanation-of-infanticide/#christian-books-3.

[13]Joseph Fielding Smith, *Man: His Origin and Destiny* (Salt Lake City: Deseret Book Company, 1965), 276.

Chapter 11

If Not Theistic Evolution, Then What?

(This chapter adapted from the author's book, *Earth In the Beginning*)

Eric N. Skousen

And the Gods organized the earth to bring forth the beasts . . . and every thing that creepeth upon the earth. – Abraham 4:25

The Remarkable Adaptive Capacity of Life Forms – Microevolution

It is well established among modern life forms that when a community of organisms is placed in a new habitat or environment, there is often a change in the appearance of subsequent generations. This is the principle of "adaptation," and is commonly observed among many different creatures (though not all). At the microscopic level we now see the rise of antibiotic resistant bacteria which are no longer vanquished with penicillin and other life-saving antibiotic medicines. They have either "adapted" to fend off the antibiotic intrusion or the bacteria which survived had some resistance—which was passed on to their descendants.

Similarly, groups of humans living and working at extremely high altitudes have developed, over many generations, larger lung capacities, increased breathing rates and dilated blood vessels to increase the distribution of oxygen through their bodies. That is, they have adapted to those conditions.

The fur of beach mice, if relocated, will gradually change color through several generations until it matches the color of the local sand thus making the beach mouse less observable to predators.[1] The peppered moth of southern England has changed from predominately light-coloring to dark-coloring during England's burgeoning industrial age. As smoke from factories darkened the local tree bark, the disguising refuge for the light-colored moths began to disappear. But it was not long before the light-colored moths produced darker-colored offspring until, as a group, most of them had dramatically changed to match the color of the sooty, ever-darkening bark.[2]

Not surprisingly, the opposite process can occur. Since very different looking descendants of a once-common parental stock can usually interbreed, it is possible for subsequent generations to revert to the appearance of their common ancestor. For example, those who have bred pigeons are aware of the many amazing varieties that can be produced from the common rock pigeon. Yet, left to themselves, carefully bred pigeons will interbreed until the offspring revert back to the common rock pigeon, as Darwin himself observed.[3]

For centuries it has been observed that in many cases the careful, selective breeding of plants or animals—which can be called "artificial selection"—will produce newly developed varieties or "species." Sometimes a new variety cannot interbreed with any of the older varieties, but only with itself.

These are all examples of microevolution—changes within the same kinds of life forms. However, it is important to note that in spite of these adaptations over many generations, a human is still a human, a mouse is still a mouse, a pigeon is still a pigeon, a moth is still a moth, and bacteria are still bacteria.

The Rise of Macroevolution

Nevertheless, in the 19th century these observed processes of adaptation suggested to scientists like Alfred Wallace and Charles Darwin that millions of years of microevolutionary change would result in the formation of entirely new life forms. As earth scientists carefully studied and classified the fossilized life forms in various strata, it became fashionable to presume that one kind of life form found in a deeper stratum was *ancestral* to a life form found in the strata above. That is, today's enormous variety of living things is believed to have been produced by repeated acts of "speciation," or the creation of entirely new kinds of life through "natural selection" as opposed to "artificial selection" (which is accomplished by plant and animal breeders).

The idea that new life forms can be created by Darwinian processes has come to be known as macroevolution, or more commonly, "evolution" or "organic evolution." The mechanism believed to cause this change was natural selection acting on sources of variation. But there was considerable opposition to the idea of "natural selection"—or as it came to be called, "survival of the fittest" or "reproduction of the fittest"—as no one knew of a source of variation in living things that would accomplish a change from one kind to another kind.

One Source of Variation

Soon after the discovery of radioactivity in 1896, there followed the unexpected disclosure that genetic material can be altered or "mutated" by chemical or radioactive substances so that certain traits in the offspring are affected. This is easily demonstrated in the scientific laboratory today. These discoveries were hailed as the plausible mechanisms by which "speciation by natural selection" or macroevolution could proceed in nature.

Natural selection acting on random genetic mutations is the modern synthesis of the theory of biological evolution. It has been the focus of a great deal of discussion and scientific research and forms the foundation for many scientific theories about the earth's early history, particularly the rise and development of life on the earth.

The Problem of Intermediate Forms

It has been pointed out, however, that if macroevolution actually occurred, there should be some evidence of the continuous development of life in the earth's crust, from the simple to the complex. However, despite an intensive search for confirming data, as of yet, earth scientists freely admit that "a detailed sequence of evolution is not evident in the rock record."[4]

Moreover, molecular biologists point out that there are complex microscopic structures basic to all living beings that cannot exist in a partially developed way—there are no "intermediate forms" for these fundamental and interdependent parts which could function in any other way than in their complete and fully-designed form. In other words, they could not have "evolved" at all; they must have simply appeared on the scene, fully and completely functional.[5]

Concerning intermediate forms in the rock record, Dr. Stephen Jay Gould, a paleontologist with Harvard University, is more specific:

> The extreme rarity of transitional forms in the fossil record persists as the trade secret of paleontology. The evolutionary trees [graphs of evolutionary genealogies] that adorn our textbooks have data only at the tips and nodes of their branches; *the rest is inference*, however reasonable, not the evidence of fossils.[6]

Even though further research may uncover additional evidence of continuous development from one "kind" to another "kind," the evidence will certainly be as inconclusive as it has been in the past. "There are so many workers in the field," Sir Fred Hoyle writes, "that nothing clear-cut can have been missed."[7] In fact, the fossilized strata, rather than being a reasonably continuous record of evolving life forms, is more like the record of repeated introductions of new life forms. Dr. Hoyle concludes that instead of a graduated record of evolving life forms, "the conclusion from geology is that the [life-form genealogical] lines . . . *were always separate*."[8]

The Sudden Appearance of New Things in the Rock Record

This view of the rock record—that there are no intermediate life forms—is confirmed by Dr. Frank B. Salisbury, a plant physiologist at Utah State University. He has pointed out that the eye, with its complex optical system, appears suddenly in the rock record not once, but three times. The wing appears in five independent animal systems. And although the fossils of many flowering plants have many similar features, no common "ancestry" has been identified.[9] In other words, new varieties of life suddenly appear in the rock record with no ancestral precursors. Even the deepest fossiliferous rocks reveal *fully developed* life forms. And more significant, between the disappearance of many life forms from the rock record and the appearance of their replacements, there are abrupt gaps where no development can be perceived. Dr. Salisbury concludes:

> There is no fossil record of gradual evolution of the phyla and divisions. . . . They are simply not to be found; complex forms appear suddenly in the record. . . . *no* major group has obvious fossil predecessors that inform us as to how the group came into being.[10]

Dogs and Horses Did Not Gradually Develop

The lack of gradual change in the fossil record is even true for popular "evidences" of macroevolution such as the development of the modern dog or the modern horse.

The "gradual development" of the dog was first put forth in a then-famous (but now infamous) series of dog fossils in 1876 by O. C. Marsh. But after decades of promotion, this series has been quietly dismissed since each member of the series is now known to have existed, unchanged, for lengthy periods in the rock record, only to be abruptly replaced by another member of the series.[11]

Similarly, in many textbooks, careful attention is given to the succession of fossil horses that appear to have gradually developed into the modern, single-toed horse. Concerning each of the so-called intermediate stages, Dr. Salisbury notes that,

> each of the intermediates is completely distinct from others in the series and *appears suddenly in the record.* There is no way of knowing that one of the so-called fossil horses gave rise by an evolutionary process to any other in the sequence.[12]

But don't we see evidences of one "kind" changing into another "kind" in controlled laboratory experiments? Dr. Armin J. Hill, former Dean of the College of Physical and Engineering Sciences at Brigham Young University, answers:

> Science has produced no evidence that at any time in the course of the development of life on earth has one kind of creature developed into another kind of creature. In fact, attempts to develop one species into another under carefully controlled laboratory conditions have succeeded about as well as the alchemist's attempts to turn lead into gold.
>
> We have recently succeeded in developing new species of grasses, but only by the most careful manipulation of the genes which control hereditary processes. We are, therefore, a long, long way from being able to accept the glibly stated hypotheses which seem to be taken for granted that [simple life forms developed into complex life forms].[13]

The Real Purpose of an Organism's Genetic Code

Scientists who probe the genetic structure of living things have discovered how to create new combinations of genes in the laboratory. This science of genetic engineering promises to provide a host of new biological miracles for all mankind. However, in both the laboratory and in nature, there is no evidence that one "kind" is being changed into another "kind." What is more apparent is that an unused portion of the genetic material, *which was already present* (or, in the genetic engineering laboratories, genetic material artificially introduced from another source), is being arranged into new combinations.[14]

In other words, the changes in the fur coloring of beach mice, or changes in the color of the peppered moths of England, or the unexpected appearance of strains of antibiotic-resistant bacteria, are not examples of one "kind" changing into another "kind." These changes are simply evidence of the amazing capacity of any life form to tap a marvelously wide spectrum of *available* genetic information stored within its own genetic code that God has supplied for each different "kind." This suggests that the real reason for a genetic code is to *retain* general characteristics of a "kind," and not create a new "kind." Evidence of this comes from the study of mutations, in which actual changes in the genetic code and the effects of these changes on subsequent offspring are observed.

We live in a radiation-filled environment in which our bodies are pummeled with cosmic rays and radiation from many other sources, most of them natural to our environment. These cause genetic damage continuously. What is amazing is that the molecular structure of our genes is replicated so many times (our genes contain thousands of repeated sequences) that our genetic storehouse can almost always repair itself from this continuous assault. It does this by simply copying information from an undamaged section.

Sometimes, however, the damage is permanent and we get what is called a "natural" mutation—a permanent change in the genetic structure. Even in established life forms, however, natural mutations only occur rarely. Furthermore, the number of mutations that are conveyed through successive generations are even rarer than that.[15] Yet, when mutations are successful, they either tend to recreate previously lost or suppressed genes or they are fatal to the offspring. And, even if the offspring survive, most are sterile and therefore unable to produce offspring of their own.[16]

All this suggests that instead of acting as a source for changes in "kind," the genetic system is a built-in control system that stabilizes the existing order of life throughout the biosphere of the earth and minimizes the possibility of new varieties of life being created from existing varieties.

Mistaking Similarity for Smart Engineering

One of the most powerful arguments for a theory of descent with modification or macroevolution—of gradual change from the simple to the complex—involves the obvious similarity of ancient life forms to modern life forms. Of course, nothing should preclude the Master Creators of the universe from employing the same engineering principles in designing preparatory life forms as in designing modern life forms.

The functional similarity of the four wheeled toys found in ancient Peruvian tombs and a modern automobile does not indicate that one "evolved" from the other. In the same sense, the similarity in appearance of a modern life form to a preparatory life form may suggest or imply that one evolved from the other, but it does not prove it.

If the huge varieties of life, preparatory and modern, had descended and developed from a few varieties of earlier life, biologists could create an acceptably consistent classification system of all known life forms. In fact, they haven't. In a typical textbook on life sciences we find this statement: "No satisfactory classification scheme exists. Many have been suggested; each has its advantages and its disadvantages."[17]

This fascinating but often underemphasized fact can be discovered by perusing any number of life science textbooks that deal with plant and animal classification. As Dr. E. O. Willey has written:

> Phylogenies [ancestral trees] for most organisms have not yet been reconstructed. . . . For example, no secondary school or college textbook published in the United States contains a classification that is logically consistent with the basic features of vertebrate phylogeny, well known though the vertebrates are.[18]

This failure, quietly admitted but seldom emphasized, stems from the false assumption that all living things are somehow related to each other through vast periods of time.

As we shall see, of the life forms that live or have lived on the earth, there are many "kinds" whose unrelated ancestry can only be traced to origins not of this earth. Their ancestors came from different planetary systems throughout the galaxy.

Both Spirit and Physical Organisms Were Created "After Their Kind"

The Lord has disclosed that in *both* the spirit and physical creations, all living creatures, whether plant or animal, were created and could only reproduce "after their kind." This was clearly the case during the spirit creation (Moses 2:11–12, 20–21, 24–25), during the physical preparation of the earth for modern life forms (Abraham 4:11–12, 20–21, 24–25), and after modern life arrived on earth (Genesis 6:20; 7:14; 8:17–19).

These truths obviously preclude macroevolution or an infinite gradation of life from one "kind" to another. Life is found in distinct "niches" such as "the horse," "the bear," and "cattle." There are also niches for creatures that fly, swim, or crawl. Joseph Smith taught:

> God set many signs on the earth, as well as in heaven [the spirit earth], for instance, the oaks of the forest, the fruit of the tree, the herb of the field—all bear a sign that seed hath been planted there; for it is a decree of the Lord that every tree, fruit, or herb bearing seed should bring forth after its kind, and *they cannot come forth after any other law or principle.*[19]

The doctrine that things only reproduce "after their kind" explains why new "kinds" have not been produced from any other "kind" in the laboratory.

A "Kind" is Not a Species

In our modern Bible, the word "kind" has been translated from the Hebrew word min. In its original context, "kind" is fairly well defined at the "Family" level in the taxonomic classification schemes now accepted by modern biologists. "Kind" never has been identified with "species" by either biologists or the scriptures. Therefore, within the wide latitude allowed by its genetic code, a "kind" can undergo marvelous changes. One "kind," however, has never developed into another "kind." As Dr. Armin J. Hill observed:

> We have been experimenting with the fruit fly – *drosophila melanogaster* – now for the equivalent of over thirty thousand years of human generations, using every known device to produce mutations. After all this, the experimenters have been forced to admit that though minor heritable variations may have been produced, the offspring is still a fruit fly.
>
> Apparently we have here extensive proof that the scripture which says in regard to plant and animal life that each shall bear "after its own kind" is the statement of a fundamental law of nature.[20]

The experiments with fruit flies mentioned by Dr. Hill extended over many decades during the early 20th century, beginning in 1909. The failure to produce major morphological changes so astonished the investigators that some of them, notably the geneticist Richard Goldschmidt, completely turned their backs on the idea of "gradual change through natural selection and mutation." Goldschmidt opted instead for an alternative explanation, particularly creation by an intelligent design.

Unfortunately, some geneticists who also arrived at Goldschmidt's conclusion have been rudely ostracized and essentially excommunicated from their scientific professions. They've been treated harshly because, by the mid-20th century, most professional scientists had already accepted the macroevolutionary view. Such is the religion of science![21]

From Whence Came the Preparatory Life?

The rock record suggests that simpler life forms *did* generally precede more complex life forms during the earth's preparatory development. And there is abundant evidence that the processes of mutation, natural selection, and adaptation to changing conditions occurred among the earliest life forms, the same as we now see occurring today among modern life forms. Yet if the simple life forms didn't evolve into complex life forms, how did this fossil record become established in the earth's crust?

The simplest explanation for the origin of life on the earth and the complex undertaking to prepare the earth's crust for modern life is this: When God chooses to introduce new life into a developing planetary system, he simply transplants this life from older stock that is already flourishing on other planets.

The interplanetary transplantation of life is not difficult to accept if we consider the following: God and his Son came from their glorified planet located at immense, cosmic distances from our earth to appear to the young Joseph Smith (Joseph Smith—History 1:17). The angel Moroni, a resurrected being, said he came "from the presence of God" (Joseph Smith—History 1:33). Other embodied beings travel at will from planet to planet (Doctrine and Covenants 130:5-6). Similarly, life forms such as plants and animals can also be transported across space from solar system to solar system.

Of course this interplanetary transplantation does not occur with the crude and ineffective technology we use today. Science fiction stories to the contrary, our present technology would almost ensure that astronauts traveling to the planet Mars would arrive in a "dangerously unhealthy" condition,[22] if not dead. We simply don't yet know how to do it, efficiently or safely.

But the Lord uses a method that is far more elegant and certainly more safe and efficient. Rather than enclosing a fragile life form in a spaceship, he changes the life form so it can survive the journey. In all likelihood He does this by temporarily translating it, just as he has already done for those who have lived on the earth but have been removed to a distant dwelling place such as the cities of Enoch and Shalem and their inhabitants (Moses 7:27; JST Genesis 14:32.)[23]

The True Origin of All Varieties of Life on the Earth

God is not required to create "new" life forms each time an earth is made. It is also not necessary to require increasingly complex life forms to somehow gradually change from "kind" to "kind" throughout the various stages of a planet's crustal preparation. In a much simpler and controlled procedure, the Creator brings life to an earth in whatever form or organization suits the particular needs. Transplantation is carried out numerous times during the preparatory creation epoch.

These are the "secret springs," Brigham Young taught, of all life forms that have ever existed on the earth:

> Life is here, but do [men] know the power that gave it, or the mode of its coming? Vegetation and animals, in great variety, teem upon the face of the earth, but are mankind familiar with the secret springs of their growth and existence? Man ought, in the first place, to find

> out how these things came, and who produced them. They will acknowledge at once that there never was a house built [except] upon simple and natural principles.[24]

Among these "natural principles," he said, is the Creator's capacity to bring "animals and the seeds from other planets to this world."[25]

From the perspective of an earth scientist, then, the observed gaps between different life forms in the strata are not due to a shortage of searching or a deficiency of discovery. These gaps are real. We *do* believe in a "God of the gaps" as they may be evidence of distinctly different life forms being repeatedly brought to the earth throughout the preparatory epoch of the earth's creation. Above all, they demonstrate that the development of life during the earth's preparatory period was directed, not haphazard or solely subject to the whims of natural selection, adaptation and mutation. Once they arrived, they propagated "after their kind," though in varieties available in the genetic code designed for the kind (microevolution).

Evidences for an Extraterrestrial Origin of Earth's Organisms

A cursory glance at the earth's present life forms gives little hint of their extra-terrestrial origin. This is because of the wide spectrum of genetic possibilities, and hence, capacities, that reside in individuals, thus enhancing their capacity for survival in vastly different environments. For example, the same variety of bacteria that subsists at enormous pressures at the ocean bottom ("baroduric" organisms) can also be found in the family garden. Other bacteria thrive in the total absence of oxygen, preferring instead an atmosphere of about 75% molecular hydrogen and 25% carbon dioxide, all at an ambient temperature of about 150 degrees Fahrenheit. Although appearing to be most un-earthlike, these are the methane-loving bacteria found in sewage sludge.[26]

Consider also that nearly every living animal has within its genetic code a blueprint to make *pterin*, a pigment that colors the wings of butterflies and the outer coverings of many insects. Yet in other creatures these genes are suppressed, or "unexpressed."[27] In fact, in the human race itself, some 90% to 95% of the genetic code of our chromosomes is believed to be unexpressed, suggesting a marvelously wide range of genetic possibilities, only some of which we see on this earth.[28] And as we have seen, it is precisely this marvelously wide spectrum of coded genes within each living creature that allows its offspring to comfortably adapt over several generations to completely new environments.

Genetic Capacities Foreign to the Earth

Among the potential genetic expressions made possible by the wide spectrum of coded genes that exist in living creatures, there are some that defy explanation. Some of these genetically-derived capacities could not have possibly evolved on earth.

The first example is the common fruit fly, *drosophila melanogaster*. In 1934, it was discovered that this creature's eyes are as sensitive to deep ultraviolet light (like that produced by a mercury arc lamp) as they are to visible light. But deep ultraviolet radiation from the sun is filtered out of the earth's atmosphere. Except for modern manmade sources, deep ultraviolet light does not exist on the earth. Nor is there any evidence that the early earth was exposed to deep ultraviolet light, even during the preparatory epoch of the earth's creation. In fact, prolonged exposure to such light is damaging, if not lethal, to bacteria and simple multicellular organisms.[29] Yet, here they are, successfully living on the earth with a capacity for vision they could not have achieved by any postulated "evolutionary processes."

Another example includes a large variety of legumes, including peas and beans. Within them can be found the suppressed genetic material to produce hemoglobin, the principle component of red blood cells, a major constituent of blood within every higher ordered animal. But only the action of the nitrogen-fixing bacterium rhizobium permits the legumes to express this particular set of genes and thus produce hemoglobin. There is no satisfactory explanation for this extremely improbable combination if it had developed only here on the earth.[30] It was apparently developed elsewhere and then adapted to the earth's present environment.

The last example (though there are many more) involves vacuum and radiation-resistant bacteria. In the course of investigating the possibility that microbes might survive a trip to Mars on the Viking spacecraft and its landers, some bacteria were found to not only survive but to flourish and multiply after five days in an ultra-high vacuum. Since an ultra-high vacuum has never existed naturally on the earth, where did these bacteria acquire this marvelous ability?

When radiation strikes cellular material, many sections of the DNA molecules making up the body of the genetic material are severed and sometimes destroyed. But as we have already noted, most of the genetic material is redundant; that is, the same sequences of molecules are repeated elsewhere in the genetic material. Thus, a damaged strand can be repaired, within limits.

These repair limits are fairly low for a creature as complex as a human being. But there exist radiation-resistant bacteria that can withstand damaging doses of X-rays requiring the successful and perfect repair of some 20,000 complete disconnections in their cellular DNA![31] Similarly, some varieties of amoeba and paramecium can successfully endure nearly as much radiation.[32]

Where did they obtain this capacity for self-repair? Certainly not on the earth via mechanisms postulated by the theory of organic evolution. Among other reasons, the extremely high radiation levels would have certainly precluded the proliferation and development of higher life forms by "evolutionary" means in the first place.

The theory that simple life forms produced complex life forms is nearly universally accepted in scientific circles today. Therefore, it is not popular to seek for and identify evidence of extra-terrestrial origins for earth life. Sir Fred Hoyle and N. Chandra Wickramasinghe, who have initiated some work along these lines, wrote:

> Our writings [on the evidences for extraterrestrial origins of life on earth] have been greeted with a wall of silence. The reason very likely is that every expert biologist has seen from the beginning that sooner or later the word purpose would appear, and to involve purpose is in the eyes of biologists the ultimate scientific sin.[33]

Hopefully, a generation of life scientists will arise who understand and take into account the work of God in seeding earth-like planets with life. Armed with this knowledge, they can begin to actively explore this fruitful arena of discovery and new understanding.

As Hoyle and Wickramasinghe conclude, "Victory in science comes only from moving in the direction in which nature points, no matter how unwelcome from a human point of view that direction may seem.[34]

The Origin and Development of Preparatory Life on the Earth

Concerning the origin and development of life on the physical earth, the Lord has revealed that life is not brought to the earth just once and left alone "to see what happens." This is the theory of unguided theistic evolution. Instead, as the need arose for more complex life forms, rather than waiting for them to somehow arise through blind and whimsical forces among the simpler life forms, each new and necessary organism was systematically brought here at the direction of God according to a specific timetable.

As Elder B. H. Roberts has written, these preparatory life forms were "brought from some other and older sphere, with power to propagate their kind."[35]

This does not mean that the observed mechanisms of natural selection, adaptation, and mutation associated with the theory of organic evolution did not operate in the earth's preparatory biosphere. But as Dr. Frank Salisbury has explained, even if they operated at maximum observed capacity, these mechanisms are inadequate, if not totally incapable, of producing the complex varieties of life that exist on the earth today:

> One positive mutation in a thousand mutations seems hopelessly optimistic. . . . Variability remains the Achilles' heel of the entire concept [of the gradual development of life forms]. Indeed, I believe that the findings of modern science during the past two decades [1960s and 1970s] have greatly increased the seriousness of the original problem, that the origin of variability and complexity is a greater mystery than ever.[36]

More important, from a purely scientific viewpoint, the effects of natural selection, adaptation and mutation—micro-evolution among the preparatory life forms—no matter how abundant or widespread, would be totally eclipsed by the impact of the repeated introduction of new life forms to the earth.

Life Forms from Other Planets

It is natural to wonder where all the preparatory life forms came from. But like Moses, we are largely limited to knowledge about life on this earth (Moses 1:35). For now the Lord has withheld specific details of the ancestors of our earth's modern life forms except to assure us that they do exist on other planets. And concerning the earth's preparatory period, He has revealed that there exist carefully prepared varieties of living things that flourish on other planets, but that do not now inhabit our earth.

The Lord showed some of these creatures to the Prophet Joseph Smith. Joseph Smith explained that these creatures are "beasts which had lived on another planet and not ours."[37] Anatomically, they occur in "a thousand forms," and live on "ten thousand times ten thousand earths like this—strange beasts of which we have no conception."[38] And among these "strange beasts" are, no doubt, a very large group of life forms that were assigned to help prepare this earth for modern life.

Both the scriptures and prophetic commentary indicate that the repeated transplantation of life to prepare the earth was a carefully executed procedure. Once the Gods initiated each creative phase of the earth's preparatory development, they "saw that they were [or would be] obeyed" (Abraham 4:12, 21, 25). That is, these Master Creators could openly express confidence that the transplanted life forms would accomplish their intended function, whether it was preparing mineral ores, coal and petroleum deposits, beds of limestone, or anything else designed for the modern inhabitants of earth to enjoy and use.

Theories of theistic evolution fail to acknowledge God's creative power and the process by which He prepared the earth for modern life forms. There is no "uncomfortable interface" between avowed macroevolution and the actual rock record, but a very comfortable interface once we accept that the preparation of the earth's crust for modern life forms was the carefully engineered result of constant guidance throughout the process.

The immovable rock against which theistic evolution is completely broken is simply this: All things procreate "after their kind." No exceptions.

Notes

[1]Gideon E. Nelson, et al., *Fundamental Concepts of Biology* (New York: John Wiley, 1970), 265.

[2]Gideon E. Nelson, et al., *Fundamental Concepts of Biology*, 263; also see H. B. D. Kettlewell, "Darwin's Missing Evidence," Scientific American 200, No. 3 (March 1959), 48–53. Unfortunately, in some biologists' zeal to promote this "evidence" of macroevolution, they resorted to trickery and deceit. See Judith Hooper, *Of Moths and Men* (New York: W.W. Norton, 2002).

[3]Charles Darwin, *The Origin of Species ...* (New York: Washington Square Press, 1963), 135–136.

[4]John G. Navarra, *Earth, Space, and Time* (New York: John Wiley & Sons, 1980), 105.

[5]See, for example, Michael J. Behe, *Darwin's Black Box* (New York: The Free Press, 1996) and Michael Denton, *Evolution: A Theory in Crisis* (Bethesda, Maryland: Adler and Adler, 1985).

[6]Stephen Jay Gould, *The Panda's Thumb: More Reflections in Natural History* (New York: W. W. Norton, 1978), 18, emphasis added.

[7]Fred Hoyle and Chandra Wickramasinghe, *Evolution From Space* (New York: Simon and Schuster, 1981), 82.

[8]Hoyle and Wickramasinghe, *Evolution From Space*, 82, emphasis added.

[9]See Frank B. Salisbury, *The Creation* (Salt Lake City, Utah: Deseret Book, 1976), Chapters 14 and 15.

[10]Salisbury, *The Creation*, 224.

[11]Alexander Mebane, *Darwin's Creation-Myth* (Glen Arm, Maryland: The Sourcebook Project, 1994), 18–19.

[12]Salisbury, *The Creation*, 224, emphasis added.

[13]Armin J. Hill, *Some Matters To Keep In Mind When Treating Science and Religion* (Provo, Utah: Brigham Young University Extension Publications, April 1966), 5–6.

[14]Salisbury, *The Creation*, 253.

[15]Natural mutations occur at a rate between once in 100,000 to once in 100,000,000 gene duplications; see Gideon E. Nelson, et al., 254.

[16]Salisbury, *The Creation*, 214–215.

[17]Gideon E. Nelson, et al., *Fundamental Concepts of Biology*, 361.

[18]E. O. Willey, in *1991 McGraw-Hill Yearbook of Science and Technology* (New York: McGraw-Hill, 1992), under "Classification, biological," 64.

[19]19 Joseph Smith, March 20, 1842, in *The Words of Joseph Smith*, Andrew F. Ehat and Lyndon W. Cook, eds. (Provo, Utah: Religious Studies Center, Brigham Young University, 1980), 107.

[20]Armin J. Hill, *Letters to My Missionary Son* (Provo, Utah: Brigham Young University Extension Publications, 1963), 2.

[21]Mebane, *Darwin's Creation-Myth*, 6–7.

[22]As pointed out by M.I.T.'s President Charles M. Vest in a November 1995 speech published as an insert in *Technology Today*, January 1996.

[23]Brigham Young, quoted in *Wilford Woodruff Journals* (Midvale, Utah, Signature Books, 1984, Scott G. Kenney, ed.) under date of December 21, 1859; also see September 8 and 25, 1859.

[24]Brigham Young, June 3, 1855, *Journal of Discourses* [henceforth as JD] (26 volumes, Liverpool and London, England, 1855–1886) 2:303.

[25]Brigham Young, April 20, 1856, *JD*, 3:319.

[26]Hoyle and Wickramasinghe, *Evolution From Space*, 66–67.

[27]to come

[28]Hoyle and Wickramasinghe, *Evolution From Space*, 20–21.

[29]F. E. Lutz and E. N. Grisewood, "Reactions of Drosophila to 2537 ÅRadiation," *American Museum Novitates*, No. 706, March 1934 (quoted in Hoyle and Wickramasinghe, *Evolution From Space*, 12–13).

[30]See Hoyle and Wickramasinghe, *Evolution From Space*, 13–21.

[31]Hoyle and Wickramasinghe, *Evolution From Space*, 21–22.

[32]Hoyle and Wickramasinghe, *Evolution From Space*, 43-44.

[33]to come

[34]Hoyle and Wickramasinghe, *Evolution From Space*, 11.

[35]B. H. Roberts, *The Gospel and Man's Relationship To Deity*, 11th edition (Salt Lake City, Utah: Deseret Book, 1966), 269–270.

[36]Salisbury, 215.

[37]Joseph Smith, April 2, 1843, *History of the Church of Jesus Christ of Latter-day Saints* [henceforth *History*] (7 vols, Salt Lake City: Deseret Book, 1951), 5:324.

[38]Joseph Smith, April 8, 1843, *History*, 5:343.

Chapter 12

Evidence for Design

Dave S. Collingridge

It's been over two thousand years since Alma and Korihor debated the existence of God. Korihor demanded evidence of God's existence before he would believe. Alma responded, "Thou hast had signs enough," and went on to describe natural phenomena as evidence for deity. He told Korihor that "the earth, and all things that are upon the face of it, yea, and its motion, yea, and also all the planets which move in their regular form do witness that there is a Supreme Creator" (Alma 30:44).

According to Alma, order in the natural world manifests intelligent creation.

Are similar arguments for the existence of deity just as valid today? They are. In fact, order in the natural world is more evident today than it was in Alma's time. Nowadays some of the most convincing evidence for intelligent creation is found in the living cell.

In the late 1800s when Darwinism was first growing in popularity, scientists believed the inside of the cell contained the foundations of life. They were right. They called the contents of the cell protoplasm – a term still in use today. Because the electron microscope had not yet been invented, they were unable to look inside the protoplasm and see what it contained. Without any good observations to go on, they surmised that protoplasm was a homogenous, jelly-like substance formed from reactions between common elements such as hydrogen, carbon, oxygen, and nitrogen.[1] This simple description of protoplasm suited nineteenth century Darwinists because it involved basic natural processes that fell within the limits of what evolution could produce.[2]

Cellular Sophistication

Advances in microscope technology have enabled us to peer into the inner workings of cells. We've learned that protoplasm is much more than a simple jelly-like substance. Among other things, protoplasm contains an amazing array of complex machinery called proteins. Proteins carry out sophisticated functions like regulating organs, sending and receiving messages, and catalyzing chemical reactions. Proteins are also specialized in the sense that there is not a one-size-fits-all protein for all functions. Specific proteins carry out specific functions. The functions they perform are determined by amino acids.

Proteins are created from chains of amino acids. There are about 20 different kinds of amino acids which can be ordered in many different ways. The high number of possible ordered chains allows amino acids to form thousands of different proteins, each carrying out a specific function. It is interesting to note that not just any order of amino acids will create a functional protein; amino acids must be ordered in a specific way to create a protein capable of performing a useful function.

How do amino acids "know" the proper way to assemble into chains? Does it happen by chance? No. An incredible amount of detailed information is required to properly order amino acids so that they produce functional proteins. After proteins are created, even more information is needed to instruct proteins on what to do. Where does all this detailed information come from? It comes from our cells' Deoxyribonucleic acid (DNA).

DNA is a double-strand structure that looks like a twisted ladder. The two sides of the ladder are sugar-phosphate strands. The rungs of the ladder are made up of pairs of nucleotide bases called Adenine, Thymine, Guanine, and Cytosine. Instructions on how to form proteins and direct their function comes from the way these four nucleotide bases are sequenced in our DNA.

To understand how nucleotide sequencing contains information, it may be helpful to think of binary computer code. Binary computer code is composed of long chains of zeros and ones. The chains of zeros and ones appear random, but in reality they contain information telling a computer what to do. Just as the sequencing of zeros and ones in binary code contains information, sequencing of the four nucleotide bases (Adenine, Thymine, Guanine, and Cytosine) in double-strand DNA contains information. That information instructs amino acids on how to form chains that produce functional proteins.

Although binary code and nucleotide sequences create information in a similar way, there are important differences. Compared to binary code, the sequencing of nucleotide bases is much more sophisticated. Microsoft co-founder Bill Gates observed that "DNA is like a computer program but far, far more advanced than any software ever created."[3] For instance, while computer memory storage devices have a limited "shelf life," DNA can last thousands of years.[4] Compared to the storage capacity of computer memory devices, DNA is vastly superior. Think about the last flash memory storage device that you used (e.g., a thumb drive). Compare the physical size and storage capacity of that device to one gram of DNA which, according to recent estimates, can store up to 455 million terabytes or 455 billion gigabytes of data.[5]

Where did DNA's detailed information come from? Who or what arranged the nucleotides so that they contain specific information? There are no known unguided, natural processes capable of ordering nucleotides into meaningful sequences. Evolutionists know this, which is why they claim it happened by chance.

The probability that nucleotides ordered themselves into information-rich sequences by chance is infinitesimally small. The probability "is so vanishingly small that its happening even once anywhere in the visible universe would count as a piece of exceptional good luck."[6] In fact, according to one estimate, the probability is so low it is below the boundary of probable events that could occur across the entire history of the universe.[7]

The improbability of nucleotide bases ordering themselves into information-rich sequences by chance does not matter to evolutionists. All that matters to them is it *could* have happened. Whether the probability is infinitesimally small or as good as one in a million, they tend to excitedly declare, "There's still a chance!"

There are no satisfactory scientific explanations of how DNA information evolved.[8] Information-packed DNA can no more evolve than the information in your computer can evolve. Unguided processes are incapable of creating massive amounts of detailed information. Detailed information can only come from one source: intelligence.

If information-rich DNA points to the existence of intelligent creation, you would think that evolutionists would acknowledge the need for intelligent creation. Unfortunately, many do not. In typical korihorian fashion, they continue to deny the role of intelligence in the creation even though the evidence for design stares them in the face.

Francis Crick, co-discoverer of the double helix structure of DNA, is an example of someone who denied the role of intelligence, despite evidence for design. Crick reminded his colleagues of the need "to constantly keep in mind that what they see was not designed, but rather evolved,"[9] lest they come to accept design-based arguments.

Some scholars have recognized DNA information for what it is – evidence of design by an intelligent source.[10] For example, in 1969 Dean Kenyon co-authored a book on natural attraction among amino acids. His book was popular with evolutionists because it proposed an undirected mechanism for forming amino acid chains. However, after further study, Kenyon realized that amino acids could not self-assemble into proteins; rather they required detailed instructions from DNA. Upon learning that amino acids required *large* amounts of information to assemble, Kenyon was led to conclude that the information could only come from an intelligent source.[11]

Trilobite Troubles for Darwin

One summer day in 2013, I was driving south on Highway 6 near Delta, Utah. My family and I were on a geological adventure to the Trilobite Quarry. I wanted to give my kids a fossil hunting adventure they would never forget.

After driving for a while, we turned off the highway and headed west for 20 miles on a dirt road. When we reached our destination, we saw a large teardrop-shaped hill surrounded by broken, grey rock. A small number of people were scattered across the base of the hill, searching for fossils. The hill is made from Cambrian shale, an ancient layer of sedimentary rock containing the fossilized remains of creatures that predate the dinosaurs. This particular site near Delta, Utah is famous for its abundant supply of trilobites, an extinct arthropod that scurried across the ancient seabed long ago.

Although I was eager to fill my five-gallon bucket with as many fossils as I could find, I took time to analyze the fossils. I noticed that trilobites had a variety of complex features. They appeared to have an intricate head structure and a jointed exoskeletan.

After two fun-filled hours of splitting rocks and gathering trilobite fossils, we were ready to go home. Everyone's bucket was full. Even my four-year-old daughter filled her bucket with fossilized remains. She found her fossils by searching through broken rocks tossed aside by previous fossil hunters, a common occurrence at the quarry as people find increasingly detailed specimens.

After arriving home, I searched through my fossils and located the best specimen, an unbroken two inch trilobite. I planned on giving it to my friend Marion Millett, a World War II veteran and former chairman of the BYU Geology-Geography Department. A few months earlier Marion had given my children spectacular fish fossils. I was going to thank him with a fossil find of my own.

When I handed the fossil to Marion, he smiled as he slid his fingers across the rough surface. He talked with uncanny familiarity about the Cambrian layer and its trilobites. As in times past, he amazed me with his knowledge of rocks and fossils.

He held up the fossil and said, "You know, trilobites are nature's best evidence against evolution."

I was curious. "How so?"

He went on to explain how the trilobite was a very sophisticated creature that suddenly appeared on earth with little or no evidence of gradual development from lower species. I had never heard of such a thing, so I decided to investigate. I searched the modern-day "seer stone" Google. It turns out Marion was right, and the evidence is mostly in the eyes.

I learned that trilobites "possessed the most sophisticated eye lenses ever produced."[12] Trilobite eyes were a complex assembly of hundreds of lenses, suggesting that its vision may have been superior to those of modern animals.[13] I also learned that the trilobite was a common creature, thus if evolution were true, there should be ample evidence of transitional life forms in Precambrian layers. These transitional creatures should have slightly less complex eyes and a slightly less complex skeletal structure. However, adequate transitional life forms have not been found. There are no fossilized creatures linking simple Precambrian creatures to more complex Cambrian trilobites.

Of the fossilized Precambrian creatures that have been found, most appear soft-bodied and small. While there is an ongoing debate about their true nature,[14] it is clear that these simple creatures bear little resemblance to more advanced, exoskeleton creatures found in the Cambrian layer.[15] Thus, Cambrian creatures like the trilobite represent a "quantum leap in organismal and ecological [complexity]" compared to Precambrian creatures.[16] This quantum leap in complexity (sometimes referred to as the Cambrian explosion) indicates that Cambrian creatures like the trilobite have no apparent ancestors in Precambrian layers as evolution supposes.

The absence of fossilized life forms bridging Precambrian and Cambrian creatures troubled Charles Darwin. He acknowledged the problem posed by the absence of transitional fossils in *The Origin of Species*:

> To the question why we do not find rich fossilif-

> erous deposits belonging to these assumed earliest periods prior to the Cambrian system, I can give no satisfactory answer. . . . [T]he difficulty of assigning any good reason for the absence of vast piles of strata rich in fossils beneath the Cambrian system is very great. . . . The case at present must remain inexplicable; and may be truly urged as a valid argument against the views here entertained.[17]

Evolution's problem of absent transitional life forms is an ongoing concern. In 2006, researchers reported that despite continued efforts to explain the Cambrian explosion through evolutionary processes, the issue continues to generate "recurrent confusion."[18] And as recently as 2011, the absence of ancestral transitional fossils was labelled "macroevolutionary lag." Macroevolutionary lag refers to the current lack of evidence for macroevolutionary change linking Precambrian and Cambrian life forms.[19] The sudden appearance of complex life forms long ago suggests that Cambrian creatures like the trilobite did not evolve – they were designed.

Scientific Materialism

In the face of compelling evidence for intelligent creation, many scholars oppose efforts to bring design arguments into science. One likely reason for their opposition is the long-held tradition that science only concerns itself with natural causes.

The idea that science concerns itself with natural causes is known as materialism. According to materialism, science should limit itself to the study of physical material and the forces governing that material. In short, with materialism, all that matters is matter. We see materialism's influence in the way science emphasizes the study of things that are perceived by the five senses (sight, touch, smell, taste, and sound). Even consciousness, an activity considered by many to be non-material, is reduced to electro-chemical processes in the brain.

Materialism has served science fairly well, to a point. It has kept pseudo-scientific claims at bay and created consensus among scientists on acceptable modes of inquiry. But the tenets of materialism have been taken too far. Despite evidence suggesting that intelligence was involved in the creation, design-based arguments are automatically excluded from consideration. Scientists oppose design-based arguments simply because they think they do not belong. This materialist attitude was summed up by one evolutionary biologist who said: "we cannot allow a divine foot in the door."[20] Ideas involving intelligent creation are not allowed simply because they violate materialism.

If science is concerned with uncovering truth about the world, it should be willing to go wherever the evidence takes it. Given the strong evidence for intelligent creation, scientists should be able to entertain design hypotheses without fear of reprisal from others.

Secularization of Science

In 2009, I gave a presentation before a gathering of LDS scholars at the Mormon Scholars in the Humanities conference held in Provo Canyon. In my presentation, I hoped to convince attendees that science should welcome acknowledgments of God. This idea is not new. During the 16th to 17th century Scientific Revolution, deity was acknowledged by some of the best scientific minds. Prominent individuals like Isaac Newton (1642–1726), Galileo Galilei (1564–1642), and Robert Boyle (1627–1691) showed us what it means to acknowledge the hand of the Lord. Here are a few examples from their scholarly writings.

> The most beautiful system of the sun, planets, and comets, could only proceed from the counsel and dominion of an intelligent and powerful being.[21] (Isaac Newton)
>
> May you long enjoy those blessings which are sent to you not so much from the stars as from God, their Maker and their Governor.[22] (Galileo Galilei)
>
> [P]henomena that discover admirable wisdom and manifest designs in the corporeal world do themselves afford a solid argument, both of the existence and of some of the grand attributes of God, with which the rest that properly belong to him have a necessary connection.[23] (Robert Boyle)

This sort of communication is no longer accepted in science. Today there is a sense that deity does not belong in science. How did we get from the God-friendly science of Newton, Galileo, and Boyle to the God-absent science that we have today? The change took place during a movement known as the Enlightenment.

The Enlightenment occurred during the 18th and early 19th centuries. It was an intellectual movement that promoted reason and science as the only reliable sources of truth. At the same time, it downplayed faith-based beliefs and marginalized religious influences. Thus, the Enlightenment transformed science into a purely secular endeavor that excluded deity from scientific discourse, a tradition that continues to this day.

Historians Palmer and Colton described the Enlightenment's secularization movement as one in which "The older

Christian view seemed no longer to be necessary." As a result, "Thinkers provided [scientific] theories of society, world history, [and] human destiny . . . in which Christian explanations had no part."[24]

The secularization of science did not go unnoticed by people living back then. William Gladstone (1809–1898) was a deeply religious, influential prime minister of England. Having noticed science's growing secularism and the threat it posed to other disciplines, he urged: "Let the scientific men stick to their science and leave philosophy and religion to the poets, philosophers, and theologian."[25]

Queen Victoria of England (1819–1901) also issued a warning. She stated, "Science is to be greatly admired and encouraged, but if it is to take the place of our Creator . . . I call it a great evil instead of a great blessing."[26]

President Joseph F. Smith (1838-1918) cautioned against the consequences of excluding deity from scientific discourse. He warned:

> In all the great discoveries in science, in the arts, in mechanics, and in all material advancement of the age, the world says, "We have done it." The individual says, "I have done it," and he gives no honor or credit to God. Now, I read in the revelations through Joseph Smith, the prophet, that because of this, God is not pleased with the inhabitants of the earth, but is angry with them because they will not acknowledge his hand in all things.[27]

My Mormon Scholars in the Humanities presentation on returning to a God-friendly science was not well received by some of the conference attendees. Their concerns were not so much pragmatic (i.e., they did not believe it was impossible to make science God-friendly). Their concerns were mostly paradigmatic (i.e., they believed science is supposed to exclude deity). This experience led me to conclude that materialism is alive and well among Latter-day Saint scholars. From a young age we've been led to believe that deity has no place in science, and this is what most of us continue to accept.

This same materialism is fueling an unusual excitement over Darwinism among some Latter-day Saints. Because Darwinism is recognized as the only legitimate scientific explanation for the origins of species, and it is accepted as truth by a large part of the scientific community, several members of the Church accept it as such. We see evidence of this in the way Darwinism is celebrated (e.g., see discussion on Darwin Week in the Preface) and defended by some members of the Church, despite its differences with the restored gospel.

Conclusion

There is abundant evidence for design in nature. Three sources of evidence covered in this chapter are the sudden appearance of diverse and complex life forms during the Cambrian era, the existence of complex machinery in living cells, and the presence of vast amounts of detailed information encoded in DNA.

The Apostle Paul taught the principle of intelligent creation in his epistle to the Romans. He wrote that ever since the creation of the world, the power of the Lord has been clearly manifested in things that are made (Romans 1:20). The food, rains, and other good things from the earth that fill our hearts with gladness bear witness of a divine Creator (Acts 14:17). So plentiful is the evidence for intelligent creation that Paul said those who choose not to believe "are without excuse" (Romans 1:20).

Paul's teachings are as relevant today as they were 2000 years ago. Evolutionists who continue to deny the influence of a designer, even though evidence for design stares them in the face, are without excuse.

Notes

[1]Thomas H. Huxley, "On the Physical Basis of Life," *Fortnight Review*, 5 (1869), available at http://aleph0.clarku.edu/huxley/CE1/Phy-sB.html.

[2]Stephen C. Meyer, *Signature in the Cell: DNA and the Evidence for Intelligent Design* (New York: Harper Collins, 2009), Chapter 1.

[3]Bill Gates, Nathan Myhrvold, and Peter Rinearson, *The Road Ahead: Completely Revised and Up-to-Date* (New York: Penguin Group, 1996), 228.

[4]Svante Paabo et al., "Genetic Analyses from Ancient DNA," *Annual Review of Genetics*, 38 (2004), 645-679.

[5]George M. Church, Yuan Gao, and Sriram Kosuri, "Next Generation Digital Information Storage in DNA," *Science*, 28 (2012), 1628.

[6]Robert Shapiro, "A Simpler Origin for Life," *Scientific American*, June 2007, 46-53.

[7]William A. Dembski, *The Design Inference: Eliminating Chance through Small Probabilities* (Cambridge University Press, 1998).

[8]Meyer, *Signature in the Cell.*

[9]Francis Crick, *What Mad Pursuit: A Personal View of Scientific Discovery* (New York: Basic Books, 1988), 138.

[10]Charles B. Thaxton, et al., *The Mystery of Life's Origin: Reassessing Current Theories* (New York: Philosophical Library, 1984), Epilogue.

[11]Meyer, *Signature in the Cell*, 237, 397-398).

[12]Lisa J. Shawver, "Trilobite Eyes: An Impressive Feat of Early Evolution," *Science News*, 105 (1974), 72–73.

[13]Shawyer, *Trilobite Eyes*, 72–73.

[14]Erwin Douglas, James Valentine, and David Jablonski, "The Origin of Animal Body Plans," *American Scientist*, 85 (1997), 126–137.

[15]Gregory Retallack, "Ediacaran Life on Land," *Nature*, 493 (2013), 89–92.

[16]Kevin Peterson et al., "The Ediacaran Emergence of Bilaterians: Congruence between the Genetic and the Geological Fossil Records," *Philosophical Transactions of the Royal Society*, 363 (2008), 1435–1443.

[17]Charles Darwin, *The Origin of Species*, 6th Edition (1872), Chapter 10.

[18]Simon C. Morris, "Darwin's Dilemma, The Realities of the Cambrian Explosion," *Philosophical Transactions of the Royal Society – Biology*, 361 (2006), 1069–1083.

[19]Erwin H. Douglas et al., "The Cambrian Conundrum: Early Divergence and Later Ecological Success in the Early History of Animals," *Science*, 334 (2011), 1091–1097

[20]Richard Lewontin, "Billions and Billions of Demons". *New York Review of Books*, January 9, 1997. Available at http://www.drjbloom.com/Public%20files/Lewontin_Review.htm

[21]Isaac Newton, *The Principia: Mathematical Principles of Natural Philosophy* (New York: Snowball Publishing, 2010, original work published in 1687), 440.

[22]Galileo Galilei, *The Starry Messenger*, Preamble.

[23]Robert Boyle, *A Free Inquiry into the Vulgarly Received Notion of Nature* (New York: Cambridge University Press, 1996, original work published in 1686.), 104.

[24]R. R. Palmer, Joel Colton, and Lloyd Kramer, *A History of the Modern World* (New York: Alfred A. Knopf, 1965), 299.

[25]Brian Silver, *The Ascent of Science* (New York: Oxford University Press, 1998), 130.

[26]Silver, *The Ascent of Science*, 130.

[27]Joseph F. Smith, *Gospel Doctrine: Selections from the Sermons and Writings of Joseph F. Smith.* Compiled by John A. Widtsoe (Salt Lake City: Deseret Book Co., 1939), 270.

Chapter 13

Questions and Answers

Q1. What is theistic evolution?
A1. Combining religious belief with the theory of evolution. (Pages 71, 159-160)

Q2. Did Darwin believe in God?
A2. Yes, but his faith in divine creation waned as he developed his ideas on the origins of species. (Pages 11-13)

Q3. Does evolution contain falsehoods?
A3. Yes. Like any other scientific theory, evolution is not perfect. Ideally we should study evolution and any other theory of science within the light of the Spirit. With the Spirit as our guide, we are less likely to fall prey to falsehoods. (Pages 14-15)

Q4. Can studying evolution cause spiritual harm?
A4. Spiritual harm can arise when people abandon or modify their religious beliefs to make them fit with evolution. (Page 4)

Q5. Does theistic evolution effectively harmonize religious and Darwinian accounts of the creation?
A5. There are aspects of evolution that cannot be harmo-

nized with gospel teachings on the creation. (Pages 71-72)

Q6. Does evolution have the stamp of approval by Church Authorities because it is taught at BYU?
A6. No. Merely because a theory is taught at BYU does not indicate the Church's endorsement. In fact, the prophets have spoken out against evolution repeatedly despite it being taught at BYU. (Pages 15, 86)

Q7. What is microevolution?
A7. Microevolution refers to changes within a species. Microevolution has been shown to take place. (Pages 5, 57, 160-161)

Q8. What is macroevolution?
A8. Macroevolution refers to changes across life forms. Macroevolution has not been shown to take place. (Pages 57, 161)

Q9. Do science and religion ever agree?
A9. Yes, truths found in the gospel and science will be in harmony. Whatever is truth is part of the gospel. (Pages 6, 20-21)

Q10. What is the difference between scientific facts and scientific theories?
A10. Facts are agreed upon evidences and observations. Theories are tentative explanations that try to make sense of collected facts. The certainty of facts should not be confused with the uncertainty of theories and inferences derived from the facts (although it often is). (Pages 23-26)

Q11. Are scientists and students being pressured to accept the theory of evolution?

A11. Scientists and students are being pressured to accept macroevolution. (Pages 6, 30-31)

Q12. What is scientific dogma?
A12. Attempts to stamp out ideas and voices opposed to evolution. (Pages 29-31)

Q13. What evidence is used to support macroevolution?
A13. Fossilization patterns, microevolution, vestigial structures, and biological similarities. (Pages 41-48)

Q14. Has macroevolution been proven beyond a reasonable doubt?
A14. Typically only in the minds of those who accept it as absolute truth. For others who reject evolutionary explanations on the origins of life, the evidence is considerably lacking. (Page 52)

Q15. What are some things evolution cannot explain?
A15. The absence of transitional fossilized creatures linking simple Precambrian and complex Cambrian life forms. Also, evolution cannot adequately explain the origins of information in DNA. (Pages 162-163, 184-190)

Q16. Why won't science consider evidence suggesting that all life was created by an Intelligent Designer?
A16. Because of materialism and the strongly held view that God does not belong in science. (Pages 190-191)

Q17. Was science ever God-friendly?
A17. Yes, during the Scientific Revolution great scientists like Galileo, Newton, and Boyle referred to deity in their scientific writings. (Pages 191-192)

Q18. When the gospel and science disagree, which one is correct?
A18. The fundamental truths of the gospel will always be correct. Scientific ideas are continually subject to change and refutation. (Pages 6, 21)

Q19. Has some part of Darwin's ideas been proven true?
A19. Yes, microevolutionary change has been proven true. (Pages 5, 56-58, 159-161)

Q20. Do biologists acknowledge the differences between micro and macroevolution?
A20. Generally speaking they avoid distinctions between micro and macroevolutionary change. They generally suppose that macroevolution is just microevolution over a long time. (Page 57)

Q21. Do objects spontaneously evolve from the simple to the complex?
A21. Spontaneous change from the simple to complex is contrary to the Second Law of Thermodynamics and seems to be opposite of the true order of natural processes. (Pages 60-61)

Q22. Have some proponents of evolution attempted to influence people's religious beliefs?
A22. Yes. Some atheists preach against religion and the existence of deity. (Pages 10, 61-65)

Q23. Have the prophets taught a consistent message with regard to evolution?
A23. Yes. The prophets have consistently denounced

macroevolution. (Pages 83-84)

Q24. Did the Prophet Joseph Smith teach anything relevant to the theory of evolution?
A24. Yes. While the Prophet died before the theory was introduced, he taught some significant principles that undermine the fundamentals of that theory. (Pages 84-85, 169)

Q25. Has a First Presidency ever issued an official statement espousing gospel principles that are contrary to the principles of evolution?
A25. Yes. The First Presidency's 1909 statement titled the "Origin of Man." A later First Presidency issued "The 'Mormon' View of Evolution" in 1925. (Pages 89-94)

Q26. What two apostles are mistakenly thought to have supported evolution?
A26. James E. Talmage and John A. Widtsoe. (Pages 97-100)

Q27. What President of the Church is mistakenly believed to have supported evolution?
A27. David O. McKay. (Pages 103-107)

Q28. Which prophet wrote a book that vehemently teaches against the principles of evolution?
A28. Joseph Fielding Smith in his book *Man, His Origin and Destiny.* (Pages 15, 115-116)

Q29. Which prophets and apostles have recently spoken and/or written out against evolution?
A29. Boyd K. Packer, Russell M. Nelson, and D. Todd Christofferson. (Pages 120-123)

Q30. Which prophets have acknowledged that macroevolution is true?
A30. None have. (Page 123)

Q31. Can evolution be completely reconciled with revealed truth found in the scriptures?
A31. No. They comprise different worldviews on the creation. (Pages 72, 127-128)

Q32. Do evolution and the scriptures disagree about the history of the earth?
A32. Very much. Evolution and the gospel disagree as to the paradisiacal earth. (Pages 128-129)

Q33. Do most evolutionists acknowledge Adam as the first man?
A33. No. According to them, evolution teaches that the first man lived long before Adam. If Adam lived, he was just one man in a long lineage of humans who came before him. (Pages 136-137)

Q34. Do evolution and the scriptures disagree with reference to the state in which man was created?
A34. Yes. Evolution states that man shares a common ancestor with animals, as a single-celled organism. The gospel teaches that man was created in the beginning in God's image. (Pages 4, 132-133)

Q35. Is there a conflict between the scriptures and evolution with reference to the timeline of human history?
A35. According to the gospel account, the first man (Adam) appeared about 6,000 years ago. According to evolution, the first man appeared long before that. (Pages 138-139)

Q36. Does evolution contradict the doctrine of no death before the Fall?
A36. Yes. According to the theory of evolution, humans were living and dying long before the time of Adam. (Pages 146-147)

Q37. How does evolution undermine the most basic and important principle of our religion?
A37. The most important principle of our religion is the Atonement of Jesus Christ. The necessity for the Atonement is based on the Fall of Adam, yet evolution rejects the Fall. (Pages 139-140)

Q38. How can the theory of evolution be spiritually destructive for some people?
A38. By making them think that they are no different than the beasts of the field, and that God had either no part, or an insignificant part in their creation. (Pages 4, 154-156)

Q39. What fundamental gospel truth about procreation contravenes evolution?
A39. All life is commanded to bring forth new life "after their kind." (Pages 148-150, 169-170)

Q40. What Biblical event contradicts the ideals of evolution?
A40. The Flood. (Pages 150-154)

Q41. If all the various living creatures did not evolve, how did they get here?
A41. Modern prophets have taught that they were brought here from other worlds. (Pages 129-131, 171-179)

Q42. Macroevolution assumes that there were intermediate life forms. What is one problem with this idea?
A42. In some cases, evidence of intermediate life forms linking one species to another as Darwin supposed is seriously lacking. (Pages 162-163, 189-190)

Q43. What modern scientific viewpoint rejects intelligent creation?
A43. Materialism. (Pages 62, 190-191)

Q44. What is one little known purpose of an organism's genetic code?
A44. To retain the general characteristics of an organism, not to create a new kind of organism. (Pages 166-167)

Q45. Does the command to create after one's kind apply to just physical creatures?
A45. No, it also applies to an organism's spiritual creation. (Pages 131-132, 169)

Q46. What may account for the fossilization pattern of creatures in the rock record?
A46. The order in which preparatory life forms were brought to earth by God. (Pages 45-46, 171-172)

Author Biographies

Dave S. Collingridge, PhD earned a Bachelor of Arts degree in psychology from the University of Lethbridge, Alberta. He earned a master's degree and PhD from Brigham Young University's Department of Psychology, Theory and Philosophy program where he studied statistics and the philosophy and history of science.

He taught statistics and research methods as an adjunct faculty member for ten years at BYU. He is currently a senior research statistician for Intermountain Healthcare where he works with clinicians researching ways to improve patient outcomes. He has published numerous peer-reviewed articles in social science and medical journals, and is the author of *Truth and Science: An LDS Perspective*, a book detailing how the Light of Christ inspires scientific discovery. He operates the website mormonsandscience.com where he blogs on science and religion and provides statistical resources to researchers. He lives in Utah with his wife and five children.

Ryan Nilsson, CPA was born and raised in Raymond, Alberta, Canada. He developed a fascination with studying the gospel which accelerated late in his teens and would often peruse his parent's library for resources on gospel learning. He distinctly remembers wondering about the apparent contradictions between science and the gospel. In the midst of this he was impressed after reading the "Evolution" entry in McConkie's classic *Mormon Doctrine*, and became convinced that no reconciliation was possible. It was not long after that he happened to stumble upon Creation Science educational material, which appeased his scientific questions and confirmed his belief in the scriptures. Soon after he served a church mission in Dallas, Texas and there was able to further reinforce his scriptural knowledge.

Ryan graduated from the University of Lethbridge in Lethbridge, Alberta, in 2011 with a Bachelor of Management, Major in Accounting – Coop Designation. Ryan currently lives in Calgary, Alberta, Canada and works as an accountant in the oil & gas industry. He maintains a website called "Loyal to the Word," (loyaltotheword.synthasite.com) which takes a conservative approach to the gospel, adhering to the scriptures and the teachings of the prophets. He has been married to his wife Amanda since his birthday in 2005 and is the loving father of four darling children.

David M. Barker, author of *Science and Religion: Reconciling the Conflicts*, has had a passion for the subject since his youth. As a retired CPA/Bank examiner he has used skills learned in his career to question popular theories and dogmas—particularly those in conflict with the scriptures. He has been especially concerned that a large number of people attribute their loss of faith to science (or more accurately stated: to scientific theories), and specifically to evolution. He graduated from Arizona State University with a BS in Business, and has also taken classes in Biology, Archaeology, Astronomy, Chemistry, Physics, The Talmud, and Communications.

He has taught numerous classes and lectures on various subjects including: conflicts between science and religion; early scriptural texts; gospel subjects; avoiding frauds, scams, and other financial problems; getting and staying out of debt; and personal merit badge management for Boy Scouts of America.

He and his wife of 43 years live in Utah and enjoy their 4 children, sons and daughters-in-law, and their 15 grandchildren.

Eric N. Skousen, PhD was born in 1941 in Washington, D.C. and was educated in California and Utah. He graduated from Brigham Young University with a Bachelor of Science degree, a Masters of Arts degree in Physics, and a Ph.D. in Mechanical Engineering Science. In addition to an extensive university teaching career and membership in scientific honor societies, Dr. Skousen has worked as an engineering scientist in the fields of energy systems, fluid dynamics, and space physics.

He is the author of several popular books on energy policy and nuclear power. He is also the author of *Earth: In the Beginning*. Widely traveled, he has lectured extensively in the United States, South America, Europe and the Middle East on gospel themes and his scientific pursuits. Dr. Skousen married Cheryl Jones of Salt Lake City, Utah and they are the parents of eight children.

Index